GARDENING FOR PROFIT;

A GUIDE TO THE SUCCESSFUL CULTIVATION

OF THE

MARKET AND FAMILY GARDEN.

ILLUSTRATED.

BY

PETER HENDERSON,

SOUTH BERGEN, N. J.

NEW-YORK:
ORANGE JUDD & COMPANY,
No. 245 BROADWAY.

Geo. W. Langford
Barber & Studio
Office 636 New York
Washington Ave
D.C.

LOVEJOY & SON,
ELECTROTYPERS AND STEREOTYPERS,
15 Vandewater street N. Y

Chas. Stout

CONTENTS AND INDEX.

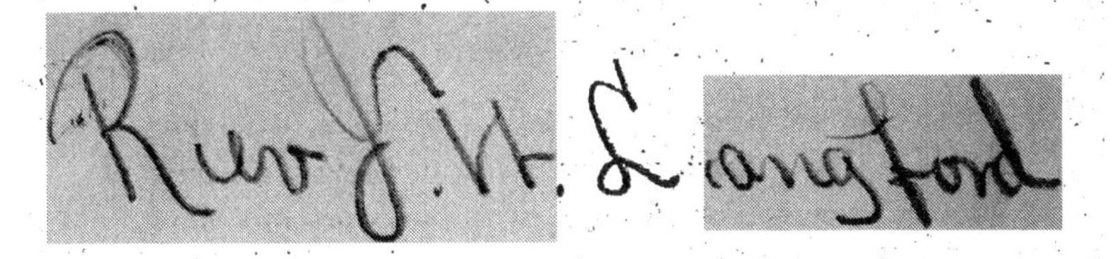

INTRODUCTION.

I hope it is no egotism to state that in both the Floral and Vegetable departments of Horticulture, in which I have been engaged for the past eighteen years, I have been eminently successful. Now, we know, that success only is the test of good generalship, and it follows that, having been successful, I have thus earned my title to merit. From this standpoint, I claim the right to attempt the instruction of the student of horticulture in the tactics of that field.

We have very few works, either agricultural or horticultural, by American authors, whose writers are practical men, and fewer still of these who are men that have "risen from the ranks." The majority of such authors being ex-editors, lawyers, merchants, etc., men of means and education, who, engaging in the business as a pastime, in a year or two generously conclude to give the public the benefit of their experience — an experience, perhaps, that has been confined to a city lot, when the teachings were of the garden, or of a few acres in the suburbs, when the teachings were of the farm.

The practical farmer or gardener readily detects the ring of this spurious metal, and excusably looks upon all such instructors with contempt. To this cause, perhaps more than any other, may be attributed the wide-spread prejudice against book-farming and book-gardening, by which thousands shut themselves off from information, the pospossession of which might save years of useless toil and privation.

I have some pride, under present circumstances, in saying, that I have had a *working experience* in all departments of gardening, from my earliest boyhood, and even to-day am far more at home in its manual operations than its literature, and have only been induced to write the following pages at the repeated solicitations of friends and correspondents, to whose inquiries relative to commercial gardening, my-time will no longer allow me to reply individually. The work has been hurriedly written, at intervals snatched from the time which legitimately belonged to my business, and therefore its text is likely to be very imperfect. I have endeavored, however, to be as concise and clear as possible, avoiding all abstruse or theoretical questions, which too often serve only to confuse and dishearten the man who seeks only for the instruction that shall enable him to practice.

Although the directions given are mainly for the market garden, or for operations on a large scale, yet the amateur or private gardener will find no difficulty in modifying them to suit the smallest requirements. The commercial gardener, from the keen competition, ever going on in the vicinity of large cities, is, in his operations, taxed to his utmost ingenuity to get at the most expeditious and economical methods to produce the finest crops—methods, that we believe to be superior to those in general use in private gardens, and which may, with profit, be followed.

Our estimates of labor, I trust, will not be overlooked; for, I know, it is no uncommon thing for gentlemen to expect their gardeners to do impossibilities in this way. The private garden cannot be properly cropped and cared for with less labor than can our market gardens, and these, we know, require nearly the labor of one man to an acre, and that too, with every labor-saving arrangement in practice. When the care of green-houses, or graperies, is in addition to this, extra labor must be given accordingly, or something must suffer.

The greatest difficulty that has presented itself to me in giving the directions for operations, has been the dates; in a country having such an area and diversity of temperature as ours, directions could not well be given for the extremes, so as the best thing to be done under the circumstances, I have taken the latitude of New York as a basis, and my readers must modify my instructions to suit their locality. The number of varieties of each vegetable described here, is very small in comparison with those that are known, or the seeds of which are offered for sale. I have given only such, as I have found most serviceable. Those who wish for a more extended list are referred to the excellent work by Fearing Burr, Jr., on Garden Vegetables.

South Bergen, N. J.,
December 1st, 1866.

GARDENING FOR PROFIT.

CHAPTER I.

THE MEN FITTED FOR THE BUSINESS OF GARDENING.

Although we shall here show the business of gardening to be a profitable one, let no man deceive himself by supposing that these profits are attainable without steady *personal* application.

Having been long known as extensively engaged in the business, I am applied to by scores every season, asking how they can make their lands available for garden purposes. The majority of these are city merchants, who for investment, or in anticipation of a rural retreat in the autumn of their days, have purchased a country place, and in the mean time they wish to make it pay; they have read or heard that market gardening is profitable, and they think it an easy matter to hire a gardener to work the place, while they attend their own mercantile duties as before. They are usually gentlemen of horticultural tendencies, read all the magazines and books on the subject, and from the knowledge thus obtained, plume themselves with the conceit that they are able to guide the machine.

Many hundreds from our large cities delude themselves in this way every season, in different departments of horticulture; perhaps more in the culture of fruits than of vegetables. I have no doubt that thousands of acres are annually planted, that in three years afterwards are abandoned, and the golden dreams of these sanguine gentlemen forever dissipated. Although the workers of the soil will not, as a class, compare in intelligence with the mercantile men of the cities, it is a mistake to suppose that this want of education or intelligence is much of a drawback, when it comes to cultivating strawberries or cabbages. True, the untutored mind does not so readily comprehend theoretical or scientific knowledge, but for that very reason it becomes more thoroughly practical; and I must say that, as far as my experience has gone, (without being thought for a moment to derrogate against the utility of a true scientific knowledge in all matters pertaining to the soil), that any common laborer, with ordinary sagacity, and twelve months' practical working in a garden, would have a far better chance of success, other things being equal, than another without the practice, even if he had all the writings, from Liebig's down, at his fingers' ends. Not that a life long practice is absolutely necessary to success, for I can see, from where I write, the homes at least of half a dozen men, all now well to do in the world, not one of whom had any knowledge of gardening, either practical or theoretical, when they started the business; but they were all *active working* men, "actual settlers," and depended alone on their own heads and hands for success, and not on the doubtful judgment and industry of a hired gardener, who had no further interest in the work than his monthly salary.

The business of market gardening, though pleasant, healthful, and profitable, is a laborious one, from which any one, not accustomed to manual labor, would quickly shrink. The labor is not what may be termed heavy, but the hours are long; not less than an average of 12 hours a day, winter and summer. No one should begin it after passing the meridian of life; neither is it fitted for men of weak or feeble physical organization, for it is emphatically a business in which one has to rough it; in summer planting, when it is of the utmost importance to get the plants in when raining, we repeatedly work for hours in drenching rains, and woe be to the "boss," or foreman, who would superintend the operation under the protection of an umbrella; he must take his chances with the rank and file, or his prestige, as a commander, is gone.

CHAPTER II.

THE AMOUNT OF CAPITAL REQUIRED, AND WORKING FORCE PER ACRE.

The small amount of capital required to begin farming operations, creates great misconception of what is necessary for commercial gardening; for, judging from the small number of acres wanted for commencing a garden, many suppose that a few hundred dollars is all sufficient for a market gardener. For want of information on this subject, hundreds have failed, after years of toil and privation. At present prices, (1866), no one would be safe to start the business of vegetable market gardening, in the manner it is carried on in the neighborhood of New York, with a capital of less than $300 per acre, for anything less than ten acres; if on a larger scale, it might not require quite so much. The first season rarely pays more than current expenses, and the capital of $300 per acre is all absorbed in horses, wagons, glass, manures, etc. If the capital be insufficient to procure these properly, the chance of success is correspondingly diminished.

I can call to mind at least a dozen cases that have occur-

r. l in my immediate neighborhood within the last five years, where steady industrious men have utterly failed, and lost every dollar they possessed, merely by attempting the business with insufficient capital. A few years ago, a man called upon me and stated that he was about to become my neighbor, that he had leased a place of twenty acres alongside of mine for ten years, for $600 per year, for the purpose of growing vegetables, and asked me what I thought of his bargain. I replied that the place was cheap enough, only I was afraid he had got too much land for that purpose, if he attempted the working of it all. I further asked him what amount of capital he had, and he told me that he had about $1000. I said that I was sorry to discourage him, but that it was better for him to know that the amount was entirely unadequate to begin with, and that there was not one chance in fifty that he would succeed, and that it would be better, even then, to relinquish the attempt; but he had paid $150 for a quarter's rent in advance, and could not be persuaded from making the attempt. The result was as I expected; he began operations in March, his little capital was almost swallowed up in the first two months, and the few crops he had put in were so inferior, that they were hardly worth sending to market. Without money to pay for help, his place got enveloped in weeds, and by September of the same year, he abandoned the undertaking.

Had the same amount of capital and the same energy been expended on three or four acres, there is hardly a doubt that success would have followed. Those who wish to live by gardening, cannot be too often told the danger of spreading over too large an area, more particularly in starting. With a small capital, two or three acres may be profitably worked; while if ten or twelve were attempted with the same amount, it would most likely result in failure. Many would suppose, that if three acres could be leased for $100 per year, that twenty acres would be cheaper at $500; nothing can be more erroneous, unless the enterprise be backed up with the necessary capital—$300 per acre. For be it known, that the rental or interest on the ground used for gardening operations is usually only about 10 per cent. of the working expenses, so that an apparently cheap rent, or cheap purchase, does not very materially affect the result. It is very different from farming operations, where often the rent or interest on purchase money amounts to nearly half the expenses.

The number of men employed throughout the year on a market garden of ten acres, within three miles of market, *planted in close crop*, averages seven; this number is varied in proportion, somewhat, according to the quantity of glass in use. I have generally employed more than that; fully a man to an acre, but that was in consequence of having in use more than the ordinary proportion of sashes. This may seem to many an unnecessary force for such a small area; but all our experience proves, that any attempt to work with less, will be unprofitable. What with the large quantity of manure indispensable, 75 tons per acre; the close planting of the crops, so that every foot will tell; the immense handling preparatory for market, to be done on a double crop each season, one marketed in mid-summer, another in fall and winter, a large and continued amount of labor is required. On lands within a short distance of market—say two miles—two horses are sufficient; but when double that distance, three are necessary. When three animals are required, it is most profitable to use a team of mules to do the plowing and heavy hauling of manure, etc., and do the marketing by a strong active horse. Every operation in cultivating the ground is done by horse labor, whenever practicable to do so; but it must be remembered that the crops of a garden are very different from those of a farm; the land is in most cases (particularly for the first crops) planted so close, that nothing will do to work with but the hoe.

CHAPTER III.

PROFITS OF MARKET GARDENING.

This is rather a difficult if not a delicate matter to touch, as the profits are so large, in some instances, as almost to exceed belief, and so trifling, under other conditions, as hardly to be worth naming. These latter conditions, however, are generally where men have started on unsuitable soils, too far from market, or without money enough to have ever got thoroughly under way. But as the object of this work is to endeavor to show how the business can be made a profitable one, I will endeavor to approximate to our *average* profits per acre. As a rule, it may be premised that for every additional acre over ten, the profits per acre will to some extent diminish, from the fact that a larger area cannot be so thoroughly worked as a smaller one; besides there will often be a loss in price by having to crowd larger quantities of produce into market, and to leave it in the hands of inexperienced salesmen· the majority of our products are quickly perishable, and must be sold when ready.

The average profits for the past fifteen years on all well cultivated market gardens in this vicinity, has certainly not been less than $300 per acre. For the past five years, (from 1861 to 1866), they have been perhaps one-third more; but these were years of "war prices," such as we will be well content never to see again. These profits are for the products of the open gardens only, not of the frames or forcing pits, which are alluded to elsewhere. These amounts are for the neighborhood of New York, which I think, from the vast competition in business, is likely to be a low average for the majority of towns and cities throughout the country. Certain it is, that from our lands, even at a value of from $1000 to $5000 per acre, we can and do profitably grow and supply the majority of towns within fifty miles around New York with fresh vegetables. In these cases, no doubt, the consumer pays full double the price that the raiser receives, for they generally pass through the hands of two classes of "middle-men," before they reach the consumer; besides which there are extra charges for packing, shipping, and freight. Thus the consumer, in a country town, where land often is not as much in value per acre as it is here per lot, pays twice the value for his partially stale vegetables or fruits, which he receives rarely sooner than twenty-four hours after they are gathered.

In most of such towns, market gardening, carried on after our manner, would, unquestionably, be highly remunerative; for if these articles were offered to the consumer fresh from the gardens, he would certainly be willing to pay more for his home-grown products, than from the bruised and battered ones that are freighted from the metropolis. Take for example the article of Celery, which pays us very well at 2 cents per root. There is hardly a city or town in the country, except New York, but where it sells for twice, and in some cases six times, that price per root;

yet the great bulk of this article sold in Philadelphia, is sent from New York, for which the consumer must pay at least double the price paid here, for it is a bulky and expensive article to pack and ship, and must of a necessity pay a profit, both to the agent here and in Philadelphia, which of course comes out of the pocket of the consumer. This is only one of many such articles of which the culture is imperfectly understood, and which the great market of New York is looked to for a supply.

The following will show the rate of receipts and expenditures for one acre of a few of the leading articles we cultivate, taking the average of the past ten years, from the grounds that have been brought up to the proper standard of fertility necessary to the market garden.

EXPENDITURES FOR ONE ACRE.

Labor	$300
Horse-labor	35
Manure, 75 tons	100
Rent	50
Seed	10
Wear and Tear of Tools, etc	10
Cost of Selling	100
	$605

RECEIPTS FOR ONE ACRE.

12,000 Early Cabbages, at 5 cts. per head	$600
14,000 Lettuce, at 1 cent per head	140
30,000 Celery, at 2 cts. per head	600
	$1340
	605
	$735

The rotation crops of Early Beets, or Onions, followed by Horseradish, or Sweet Herbs, as a second crop, give nearly the same results.

CHAPTER IV.

LOCATION, SITUATION, AND LAYING OUT.

LOCATION.—Before deciding on the spot for a garden, too much caution cannot be used in selecting the locality; mistakes in this matter are often the sole cause of want of success, even when all other conditions are favorable. It is always better to pay a rent or interest of $50 or even $100 per acre on land one or two miles from market, than to take the same quality of land, 6 or 7 miles dis[illegible] for nothing; for the extra expense of teaming, procuring manure, and often greater difficulty in obtaining labor, far more than counterbalance the difference in the rental of the land. Another great object in being near the market is, that one can thereby take advantage of the condition of prices, which often, in perishable commodities like garden produce, is very variable. It not unfrequently happens that from scarcity or an unusual demand, there will be a difference of $25 or $30 per load, even in one day, hence if near a market, larger quantities can be thrown in than if at a distance, and the advantage of higher rates be taken.

This disadvantage in distance only holds good in perishable articles, that are bulky; the lighter and valuable crops, such as Tomatoes, Cucumbers, Lettuce, Radishes, etc., from more southerly and earlier localities, are grown often hundreds of miles distant, and freighted to market at a handsome profit. So with less perishable articles, such as dry roots of Carrots, Beets, Parsnips, Horseradish, etc.; but the necessity of nearness to market for the bulky and perishable crops, is imperative.

SITUATION AND LAYING-OUT.—It is not always that choice can be made in the situation of or aspect of the ground; but whenever it can be made, a level spot should be selected, but if there be any slope, let it be to the south. Shelter is of great importance in producing early crops, and if a position can be got where the wind is broken off by woods or hills, to the north, or northwest, such a situation would be very desirable. In the absence of this, we find it necessary to protect, at least our forcing and framing grounds, with high board fences, or better yet, belts of Norway Spruce. The most convenient shape of the garden is a square or oblong form; if square, a road 12 feet wide should be made through the centre, intersected by another road of similar width, see (fig. 1); but if oblong, one road of the same width, running through the centre in a plot of ten acres, will be sufficient.

Fig. 1.—PLAN OF GROUND.

VEGETABLE HOUSE, WELLS, ETC.—Connected with every market garden is a vegetable house, usually about 25 feet square, having a frost-proof cellar, over which is the vegetable or washing house. In the second story is a loft for seeds, storage, etc. Immediately outside the vegetable house is the well, from which the water is pumped to a tub in one corner of the building, on each side of which are erected benches of convenient hight on which the workmen tie and wash the vegetables preparatory to sending them to market.

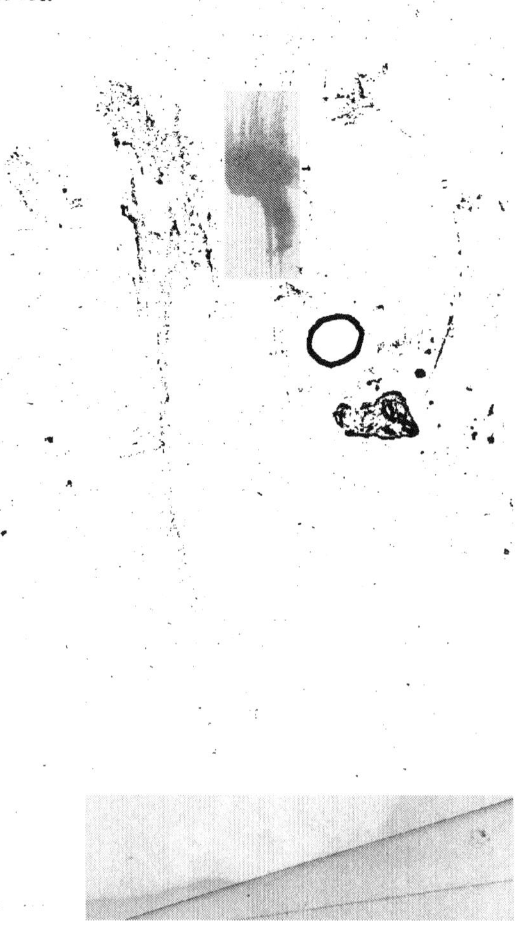

CHAPTER V.

SOILS, DRAINAGE, AND PREPARATION.

In the course of an experience of nearly 20 years as a market gardener, in the neighborhood of New York, I have had, in the prosecution of the business, the opportunity of reclaiming large tracts of very different varieties of soil. Some of these, almost the first season, yielded a handsome profit, while with others, the labor of years, and the expense of large sums in extra manuring and draining, have never been able to bring these uncongenial soils up to the proper standard of productiveness.

The variety of soil that we value above all others, is an alluvial saline deposit, rarely found over more than a mile inland from the tide mark. It is of dark heavy loam, containing, throughout, a large mixture of decomposing oyster and other shells; it averages from 10 to 30 inches deep, overlaying a subsoil of yellow sandy loam. The next best variety is somewhat lighter soil, both in color and specific gravity, from 8 to 15 inches deep, having a similar subsoil to the above. Then we have a still lighter soil, in both senses of the term, in which the sand predominates over the loam, and laying on a subsoil of pure sand; this variety of soil is well adapted for Melons, Cucumbers, Sweet Potatoes, Radishes, and Tomatoes, but is almost useless for growing crops of Onions, Cabbages, or Celery. We have still another kind of soil, which I place last, as being of the least value for the purpose of growing vegetables; this variety, singularly enough, is found on the highest points only, its color is somewhat lighter than the variety first mentioned; it is what is termed a clayey loam, averaging ten inches in depth, under which is a thick stratum of stiff bluish clay. With a subsoil of this nature, it is almost useless to attempt to grow *early* vegetables for market purposes.

I have just such a soil, as the last mentioned, thoroughly drained three feet deep, the drains only 18 feet apart, and yet, in another garden, that I work, having the two first named soils and only one mile distant, manured and cultivated the same in all respects, fruits and vegetables are ready from 5 to 10 days earlier. But for the succession, or second crops, such as Celery, etc., this stiff cold soil is just what is wanted; earliness with these is not the object, and its "coldness" is congenial to the roots of the late crop. But if selection can be made for general purposes, choose a rather dark-colored loam soil, neither "sandy" nor "clayey," as deep as can be found, but not less than 12 inches. If it overlay a sandy loam of yellowish color, through which water will pass freely, you have struck the right spot, and abundant crops can be raised under proper management. When selecting land, do not be deceived by any one who tells you, that if not naturally good, the soil may be made so by cultivation and manure. These will help, certainly; but only as education improves the shallow mind. Luxuriant crops can no more be expected from a thin and poor soil—no matter how much it is cultivated—than fertile ideas from a shallow brain, educate it as you will.

DRAINAGE.—Every operator in the soil concedes the importance of drainage, yet it is really astonishing to observe how men will work wet lands year after year, wasting annually, by loss of crops, twice the amount required to thoroughly drain. A most industrious German, in this vicinity, cultivated about 8 acres for 3 years, barely making a living; his soil was an excellent loam, but two-thirds of it was so "spongy," that he could never get it plowed until all his neighbors had their crops planted. Driving past one day, I hailed him, asking him why he was so late in getting in his crop, when he explained that if he had begun sooner, his horses would have "bogged" so, he might never have got them out again. I suggested draining, but he replied that would never pay on a leased place; he had started on a ten years lease, which had only 7 years more to run, and that he would only being improving it for his landlord, who would allow him nothing for such improvement. After some further conversation I asked him to jump into my wagon, and in 10 minutes we alighted at a market garden, that had 6 years before been just such a swamp hole as his own, but now, (the middle of May), was luxuriant with vegetation. I explained to him what its former condition had been, and that the investing of $500, in drain tiles, would, in 12 months, put his in the same condition. He, being a shrewd man, acted on the advice, and at the termination of his lease, purchased and paid for his 8 acres $12.000, the savings of six years on his drained garden. I honestly believe, that, had he gone on without draining, he would not have made $1200 in 12 years, far less $12.000 in 6 years. My friend estimates his whole success in life to our accidental meeting and conversation that May morning, and consequently I have no stauncher friend on earth than he.

The modes of draining must be guided to a great extent by circumstances; wherever stones are abundant on land, the most economical way to dispose of them, is to use them for drainage. I have also used with great success, in a wet sandy subsoil, where digging was easily done, brush, from adjacent woods cut off, and trod firmly 2 feet deep in the bottom of drains 5 feet deep, overlaying the brush with straw or meadow hay before covering in. Drains so made, have answered well for nearly a dozen years, and in situations where no other material offers, they will at least answer a temporary purpose. But unquestionably, when at all attainable, at anything like reasonable cost, the cheapest and most thorough draining is by tile. We use here the ordinary horse-shoe tile; 3-inch size for the laterals, and from 5 to 6 inch for the mains. On stiff clayey soils, we make our lateral drains 3 feet deep, and from 15 to 18 feet apart; on soils with less compact subsoils, from 20 to 25 feet distant. We find it cheaper to use the horse-shoe than the sole tile; in lieu of the sole we cut common hemlock boards in 4 pieces; that is, cut them through the middle, and split these again, making a board, thus cut, run about 50 feet; these are placed in the bottom of the drains, and prevent the sagging of the tiles in any particular spot that might be soft, (fig. 2). We are particularly careful to place, after setting, a piece of sod, grass down, over the joinings of the

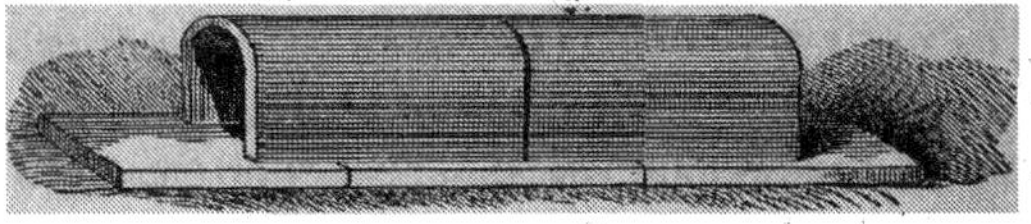

Fig. 2.—HORSE-SHOE TILE.

tiles, to prevent the soil from getting in and stopping up the drainage.

The manner of constructing stone drains, is governed by the character of the stone on hand; if round, they are best made as rubble drains, (fig. 3); but if flat, which is much the best, they are made as represented by fig. 4.

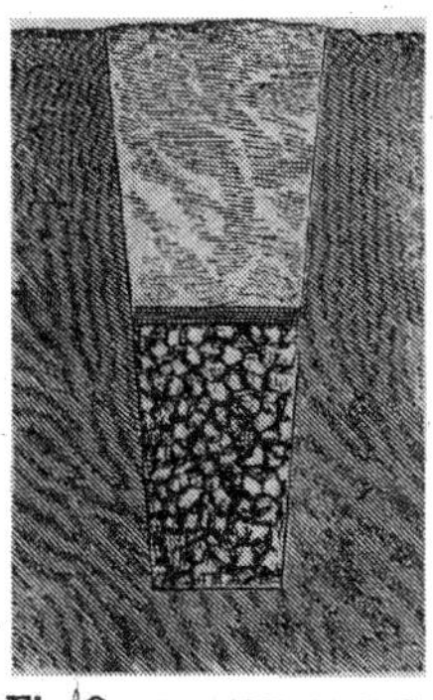

Fig. 3.—RUBBLE DRAIN.

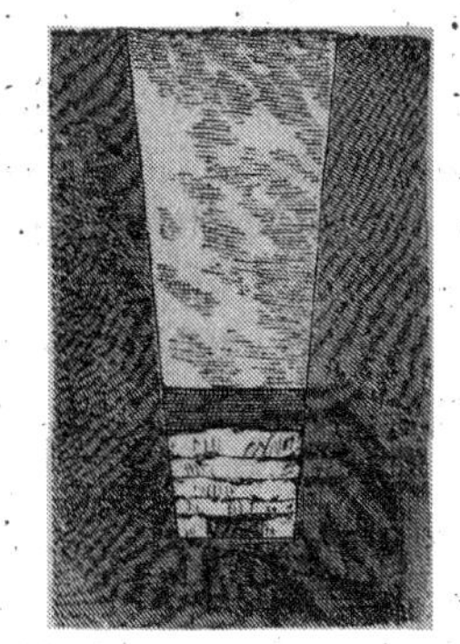

Fig. 4.—FLAT STONE DRAIN.

But in either case, the same care must be exercised in covering over the top, thoroughly, with sod, shavings, straw, or some similar material, in quantity sufficient to prevent the soil from washing in and filling up the cavity.

PREPARATION OF THE GROUND.—Assuming that the ground on which the garden is to be formed is in sod, the best time to begin operations is in September, October, or November. If draining is necessary, that should be first completed. Before the sod is plowed, it would greatly assist its rotting, if horse manure can be obtained, to spread it over the surface, to the depth of two or three inches. In plowing the sod under, care should be taken to have it laid as flat as possible; this can be best done by plowing shallow, and at this time there is no particular necessity for deep plowing. After plowing, we find it advantageous to flatten down the furrows, by running over with the *back* of the harrow; this mellows the soil so that it fills up the crevices left between the furrows, and hastens the decomposition of the sod. If the plowing has been done early enough in the fall, so that the sod has had time to rot the same season, it will facilitate the operations of next spring to cross plow and thoroughly harrow; but if too late, this had better be deferred until spring. After the ground has been well broken up by this second plowing and harrowing, it should again be manured over the whole surface with rough stable manure, as much as can well be procured; there is rarely danger of getting too much, and the third plowing takes place, followed this time by the subsoiler. I have always found it best, in breaking in new ground, to crop with Potatoes, Corn, or late Cabbages the first season,—it rarely indeed happens that any amount of labor or manuring can so prepare the ground, the first season, as to bring it to that high degree of tilth necessary for growing garden vegetables as they should be grown, and any attempt to do so, will result in a meagre crop, which will not pay—at least in such districts as New York, where there is always abundance of products of the first quality. It must not be expected that the crops of Potatoes, etc., will give much profit for this unusual outlay in preparation and manure, for they certainly will not, and the beginner must be content to wait for his profits until the second season; these are certain to be realized if these preparations have been properly made, hence it will be seen the necessity for capital in this business, for the returns, though highly remunerative, are not quick.

CHAPTER VI.

MANURES.

The quantity, quality, and proper application of manures, is of the utmost importance in all gardening operations, and few have any conception of the immense quantity necessary to produce the heavy crops seen in our market gardens. Of stable or barn-yard manure, from 50 to 100 tons per acre is used, and prepared, for at least six months previously, by thorough turning and breaking up to prevent its heating unduly. The usual method is to have the manure-yard formed in a low part of the garden, but if there is no natural depression, one may be made by digging out from 18 to 24 inches deep, and enclosing it by a fence about 6 feet in hight. The wagons are driven alongside, and the green manure thrown into the enclosure, care being taken to have it spread regularly; hogs are usually kept upon the manure in numbers sufficient to break it up, they being fed in part by the refuse vegetables and weeds of the garden.

The manure of horses is most valued, as we consider it, weight for weight, of about one-third more value than that of cows or hogs; on stiff soils it is of much more benefit as a pulverizer. There are many articles, the refuse of manufactures, that are still wasted, that have great value as manures. Among others, and of first importance, is the refuse hops from the breweries. It is a dozen years ago since they first began to be used in our gardens about New York; at first they were to be had almost at every brewery without cost, but the demand has so increased, that the price to-day ranges even higher than that of the best stable manure. Aside from its high fertilizing properties, it is excellent for breaking up and pulverizing the soil, and as a top-dressing or mulching, either to protect from the sun in summer, or from the frost in winter, it has no equal. From my experience with this fertilizer, I consider it to be of nearly double the value of that of stable manure. It requires to be composted in the same manner as other manure; it heats rapidly, and must be either spread regularly over the hog yard, or else turned once in two weeks to prevent "fire-fang," from violent heating.

Another valuable refuse from our manufactories is the shavings and scrapings from horn, or whale-bone manufactories. The best way to render these most available, is to compost them thoroughly with hot manure, in the proportion of one ton of shavings to fifteen of manure; the heated manure extracts the oil from the shavings, which is intermingled with the whole. I have on several occasions seen the mixture of five tons of whale-bone shavings with our ordinary stable manure, make $400 per acre difference in the value of the crop; but of course such manufactories are not common, and it is only in certain localities that this fertilizer can be had.

Another valuable fertilizer from manufactories is "sugar house scum," which is composed largely of blood, charcoal, and saccharine refuse; as it heats violently, instead of being thrown in heaps by itself, it should be composted with equal quantities of soil or muck, and turned frequently, so that the whole is thoroughly mixed; thus when composted, it makes an excellent manure at twenty tons per acre; it is best applied by lightly plowing, or deeply harrowing-in.

Of concentrated manures, perhaps the best for general purposes, is pure Peruvian guano; this for general crops, when used without the addition of stable manures, is put on at the rate of from 1000 to 1200 pounds per acre; it is first pounded to powder so that it can be regularly sown over the surface, after plowing; it is then thoroughly harrowed in, and the crop is sown or planted at once. In my experience, the next best concentrated fertilizer is bone-dust, or flour of bone; in experiments last season, with our crops of cauliflower and cabbage, we applied it in the same manner as guano, but at the rate of nearly 2000 pounds per acre, and it gave most satisfactory results, surpassing those of guano, where that had been used at the rate of 1200 pounds per acre. In applying manures to the soil, we have long ago discovered the great importance of an alternation of different kinds; when I first began business as a market gardener, I had opportunities of getting large quantities of night soil from the scavengers of Jersey City; this was mixed with stable manure, charcoal, sawdust, or any other absorbent most convenient, and applied so mixed at the rate of about 30 tons per acre. The crops raised with this manure were enormous, for two or three years, but it gradually began to lose effect, and in five years from the time we began to use it, it required nearly double the weight of this compost to produce even an average crop. I then abandoned the use of night soil and applied refuse hops instead, at the rate of about 60 tons per acre, with marked improvement; but this was for the first and second years only, the third year showing a falling off. About this time our prejudices against the use of concentrated manures for market gardening began to give way, and at first we applied guano together with manure at the rate of 300 pounds per acre, which we found to pay; and the next season, guano was used at the rate of 1200 lbs. per acre, with very satisfactory results. Since then, our practice has been a systematic alternation of manures, which I am convinced is of quite as much importance to the production of uniform crops of first quality, as is the alternation of varieties of the different kinds of vegetables.

It is a grave blunder to attempt to grow vegetable crops, without the use of manures of the various kinds in about the proportions I have named. I never yet saw soil of any kind that had borne a crop of vegetables that would produce as good a crop the next season without the use of manure, no matter how "rich" the soil may be thought to be. An illustration of this came under my observation last season. One of my neighbors, a market gardener of nearly twenty years' experience, and whose grounds have always been a perfect model of productiveness, had it in prospect to run a sixty-foot street through his grounds; thinking his land sufficiently rich to carry through a crop of Cabbages, without manure, he thought it useless to waste money by using guano on that portion on which the street was to be, but on each side sowed guano at the rate of 1200 pounds per acre, and planted the whole with Early Cabbages. The effect was the most marked I ever saw; that portion on which the guano had been used sold off readily at $12 per hundred, or about $1400 per acre, both price and crop being more than an average; but the portion from which the guano had been withheld, hardly averaged $3 per 100. The street occupied fully an acre of ground, so that my friend actually lost over $1000 in crop, by withholding $60 for manure. Another neighbor, whose lease had only one year to run, and who also unwisely concluded that it would be foolish to waste manure on his last crop, planted and sowed all without it; the result was, as his experience should have taught him, a crop of inferior quality in every article grown, and loss on his eight acres of probably $2000 for that season.

CHAPTER VII.

IMPLEMENTS.

The most important implements in use in the vegetable garden are the plow and harrow, which should be always used, to the exclusion of the spade or digging fork, whenever it is practicable to do so. No digging, in the ordi-

Fig. 5.—ALLEN'S CYLINDER PLOW.

nary way, can pulverize the soil so thoroughly as can be done by the plow and harrow, nor no trenching much surpass in its results that done by thorough subsoiling. Fig. 5 represents the plow in use by the market gardeners, and known as Allen's Patent Cylinder Plow. So superior are the pulverizing powers of this plow to those of the spade, that no market gardener here, of any experience, would allow his grounds to be dug, even if it were done so free of cost.

Fig. 6 represents the Lifting Subsoil Plow, it is strongly made, of steel, and follows in the wake of the surface plow,

Fig. 6.—LIFTING SUBSOIL PLOW.

lifting and breaking (but not turning) the subsoil to the depth of 6 or 12 inches, as may be desired. On very stiff soils we use the subsoiler once in two years; on lighter soils not so often, although if time would always permit there is no doubt but that it would be beneficial to use it whenever plowing is done.

The harrow in use is rather peculiar in style, but is best suited for garden work; it contains some forty teeth about 10 inches long; these are driven through the wood-work, leaving 5 or 6 inches of the sharpened end on the one side and from 1½ to 2 inches of the blunt end on the other, as shown in fig. 7. After the ground has

Fig. 7.—GARDEN HARROW.

been thoroughly pulverized by the teeth of the harrow, it is turned upside down, and "backed," as we term it; the short blunt teeth further breaking up the soil and smoothing it to a proper condition to receive the seeds or plants.

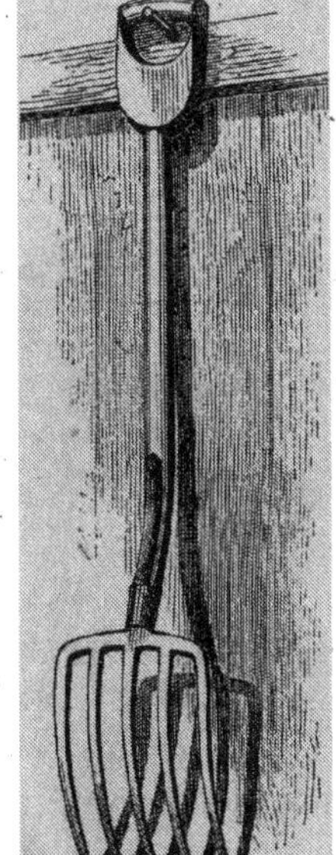
Fig. 8.—DIGGING FORK.

But there are many spots in the garden that it is impracticable to plow, such as our frames, borders, and occasionally between rows where the space is too narrow for a horse to walk; such places must be dug, and here we use the Digging Fork, represented by fig. 8, in preference to the spade. Its prongs enter the soil more easily than the blade of the spade, and by striking the soil turned over, with the back of the fork, it pulverizes it better than can be done by the blade of the spade. Still there are many operations in the garden, such as the digging up of roots, earthing up of Celery, etc., for which the spade is indispensable. For such purposes, the one represented by fig. 9, and known as "Ames' No. 2, Plain-back," we find the best.

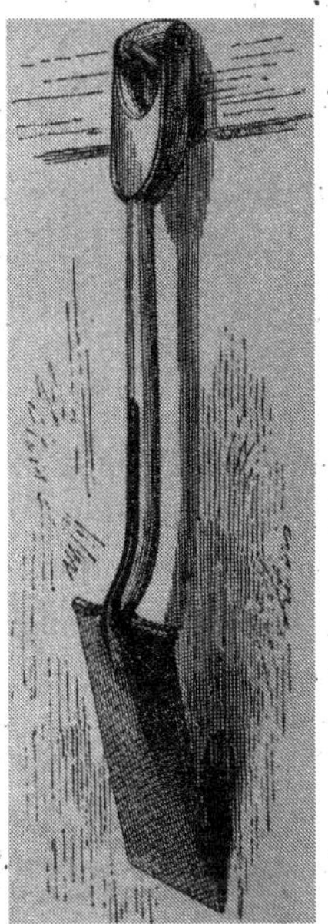
Fig. 9.—AMES' No. 2 SPADE.

For stirring between narrow rows of Cabbage, Celery, etc., we use a small one-horse plow before using the cultivator; this is represented by fig. 10, and is known as the Skeleton Plow. Following this is our main implement

Fig. 10.—SKELETON PLOW.

for cultivating between rows, which is simply a triangular adjustable Harrow, represented by fig. 11. This implement we prefer to any variety of cultivators we have ever used, on ground where there are no weeds, (and weeds are rarely allowed to grow in our market gardens), as its teeth sink from three to four inches deep if kept sharpened; when extra depth is wanted, a weight

Fig. 11.—CULTIVATOR.

is put on to sink it deeper. In all hoeing operations by hand, the steel pronged Hoe, fig. 12, is used in preference to the old-fashioned blade hoe; yet, superior as this implement is to the blade hoe, it is not more than six years ago since it came into general use. A man can do full one-third more work with it, do it better, and with greater ease, than with the blade hoe; true, it is not so good in cutting over weeds, but weeds should never be seen in a garden, for whether for pleasure or profit, it is short-sighted economy to delay the destruction of weeds until they start to grow. One man will hoe over, in one day, more ground where

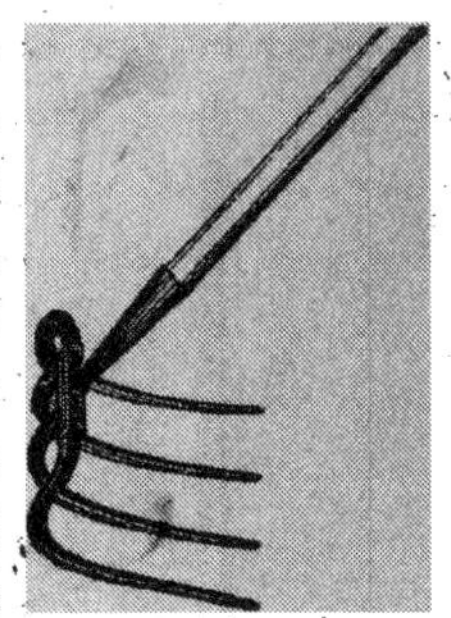
Fig. 12.—PRONGED HOE.

the weeds are just breaking through, than six will, if they be allowed to grow six or eight inches in hight, to say nothing of the injury done to the ground by feeding the weeds instead of the planted crops. Another benefit of this early extirpation of weeds is, that taken in this stage, they of course never seed, and in a few years they are almost entirely destroyed, making the clearing a much simpler task each succeeding year.

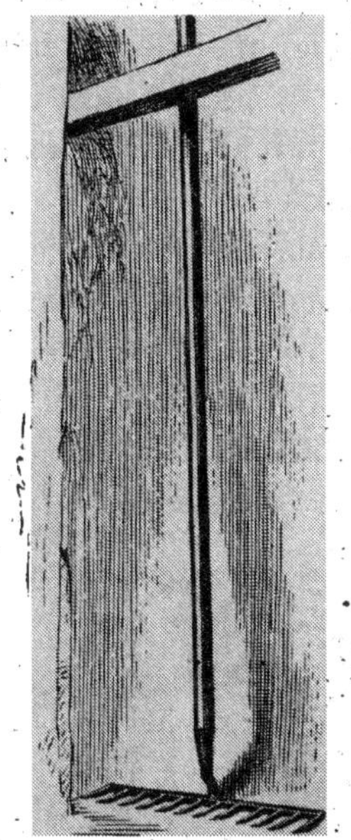
Fig. 13.—STEEL RAKE.

Another tool used in place of the hoe, is the steel Rake, fig. 13, which we use in various sizes, from 8 inches to 20 inches in width. Nearly all our first "hoeing" is done with these; that is, the ground is raked over and levelled in from two to three days after planting; this destroys the germs of the weeds; in from five to ten days, according to the state of the weather, the ground is again gone over with the rakes. We are no believers in deep hoeing on newly planted or sown crops, it is only when plants begin to grow that deep hoeing is beneficial.

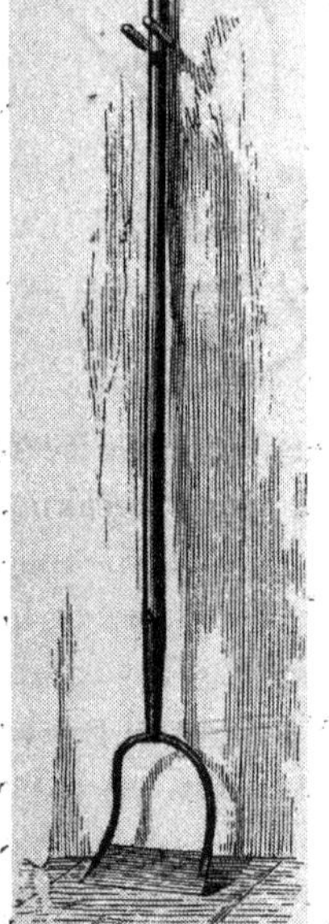

Fig. 14.—SCUFFLE HOE.

For using between narrow rows of crops, just starting from the ground, the push or Scuffle Hoe, (fig. 14), is a most effective tool; we use them from 6 to 12 inches wide; they require to be always about 3 inches narrower than the rows; thus, in rows 9 inches apart, we use the 6-inch hoe.

The Clod Crusher, fig. 15, an implement much used in England, is of great value in pulverizing the surface of rough heavy soils, following after the harrow; on light soils, that pulverize sufficiently with the harrow, it is not necessary.

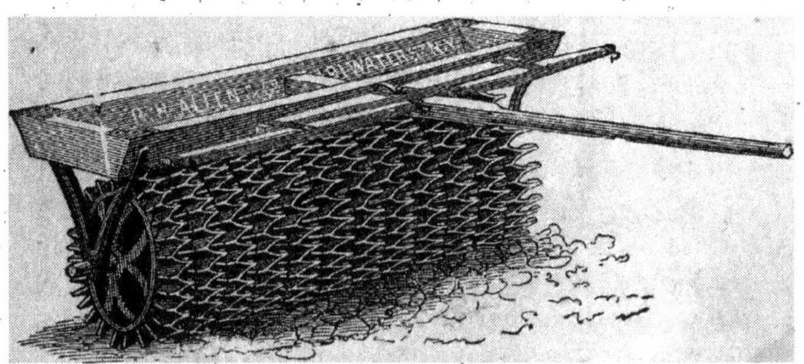

Fig. 15.—CLOD CRUSHER.

Another indispensable implement is the Roller, fig. 16; it is of great importance not only in breaking lumpy soil, but in firming it properly around newly sown seeds, besides, the ground leveled by the roller is much easier hoed than if the surface were uneven or irregular. The roller we use is made of hard wood, and is 5 or 6 feet long, and 9 inches in diameter. The roller is bored though its whole length, and through this hole is put a bar of 2-inch round iron. This bar gives the necessary weight, and its projecting ends afford points to which to attach the handle.

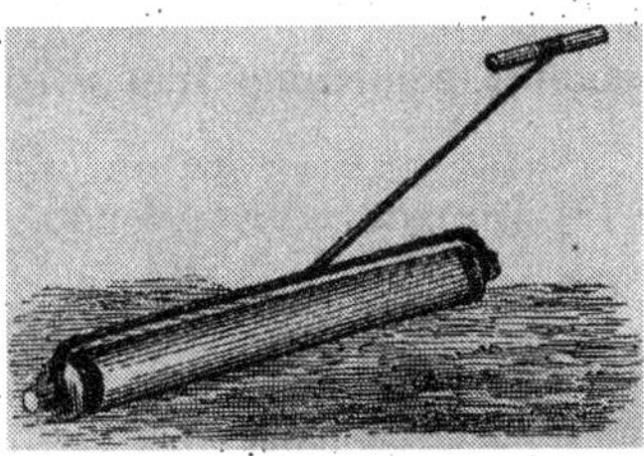

Fig. 16.—GARDEN ROLLER.

The Double Marker, fig. 17, is used to mark 6 or 8 lines at once, as may be required; the spaces between the teeth being 12 inches on one side, and 9 inches on the other. Where rows are required only of these widths, every row is of course planted, but many of our crops require wider rows, thus, with the 12-inch marker, we plant our early Cabbages at 24 inches apart, the intervening rows being planted with Lettuce at the same time; or with the narrow side of the marker, every row, 9 inches apart, is planted with Onion sets, or in such a crop as Beets, every alternate row only is used, making the rows 18 inches apart. The

Fig. 17.—DOUBLE MARKER.

manner of using the marker will readily suggest itself. A line being stretched tightly to the required length, the outer tooth is set against it and steadily drawn to the end, returning, the outer row forms the guide for the marker, and so on until finished. The marker is usually a home-made implement, of wood, but it answers rather better to have the teeth made of iron, scooped, something like a common garden trowel.

The Market Wagon (figure 18) is made after various patterns in different sections of the country; that shown

Fig. 18.—MARKET WAGON.

in the cut is the kind used by us, and is usually drawn by one horse, it is strongly made, weighing about 1400 pounds, and is capable of carrying from 2000 to 3000 pounds.

The Seed Drill, fig. 19, next page, is used in sowing large field crops of Onions, Carrots, Turnips, etc., and can be adjusted to suit all sizes of seeds. It is, however, more an implement of the farm than the garden, and rarely used in small market gardens, most cultivators deeming it safer to sow by hand. Sowing by hand requires more than twice the quantity of seed than when sown by the drill, but the crops of our market gardens are too important to run any risk from such small considerations of economy. The greater risk in thinly sown crops being from destruction by insects, frost, or the thin sowing not

Fig. 19.—WETHERSFIELD SEED DRILL.

having strength enough to force through the soil in dry weather.

The Dibber, fig. 20, is a very simple but indispensable tool. It is of importance to have it made in the manner represented here; it can be formed from a crooked piece of any hard wood, and shod with a sharp iron point, which gives weight to it, besides it always keeps sharp. Dibbers are too often made from an old spade or shovel handle, when they are awkward and unhandy affairs.

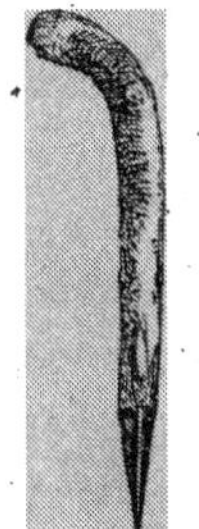

Fig. 20.—DIBBER.

Planting is an operation that often requires the most rapid movement to get a crop in at the proper time, and the best appliances in working are not to be disregarded. With a dibber of this style, an experienced planter, with a boy to drop the plants, as we invariably practise, will plant from 6000 to 10,000 plants per day, according to the kind of plant or condition of the ground. I have on many occasions planted, in one day, three acres of Celery, holding about 90,000 plants, with ten men, each of whom had a boy, from ten to fourteen years of age, to drop the plants down before him. This plan of using boys is not generally adopted, but I have repeatedly proved that, by thus dividing the labor, a boy and a man will do more planting than two men would if planting singly, and each carrying his own plants.

CHAPTER VIII.

THE USES AND MANAGEMENT OF COLD FRAMES.

We use cold frames for preserving Cauliflower, Cabbage and Lettuce plants during the winter, and the forwarding of Lettuce and Cucumbers in spring and summer.

To make the matter as clear as possible, we will suppose that the market gardener, having five or six acres of land, has provided himself with 100 of 3×6 feet sashes. The Cauliflower, Cabbage, or Lettuce plants, which they are intended to cover in winter, should be sown in the open garden from the 10th to the 20th of September, and when of sufficient size, which they will be in about a month from the time of sowing, they must be replanted in the boxes or frames, to be covered by the sashes as winter advances.

The boxes or frames we use, are simply two boards, running parallel, and nailed to posts to secure them in line. The one for the back is ten or twelve inches wide, and that for the front seven or eight inches, to give the sashes, when placed upon them, pitch enough to carry off rain, and to better catch the sun's rays. The length of the frame or box may be regulated by the position in which it is placed; a convenient length is fifty or sixty feet, requiring eighteen or twenty sashes.

Shelter from the north-west is of great importance, and if the ground is not sheltered naturally, a board fence six feet in hight is almost indispensable. The sashes should face south or south-east. Each sash will hold five hundred plants of Cabbage or Cauliflower, and about eight hundred of Lettuce. These numbers will determine the proper distance apart, for those who have not had experience. It should never be lost sight of, that these plants are almost hardy, and consequently will stand severe freezing without injury; but to insure this condition they must be treated as their nature demands; that is, that in cold weather, and even in clear winter days, when the thermometer marks 15 or 20 degrees in the shade, they must be abundantly aired, either by tilting up the sash at the back, or better still, when the day is mild, by stripping the sash clear off. By this hardening process, there is no necessity for any other covering but the sash. In our locality, we occasionally have the thermometer from 5° to 10° below zero for a day or two together, yet in all our time we have never used mats, shutters, or any covering except the glass, and I do not think we lose more than two per cent. of our plants. Some may think that the raising of plants in this manner must involve considerable trouble, but when they are informed that the Cabbage and Lettuce plants so raised and planted out in March or April, not unfrequently bring a thousand dollars per acre before the middle of July, giving us time to follow up with Celery for a second crop, it will be seen that the practice is not unprofitable.

But we have not yet done with the use of the sashes; to make them still available, *spare* boxes or frames must be made, in all respects similar to those in use for the Cabbage plants. These frames should be covered up during winter with straw or leaves in depth sufficient to keep the ground from freezing, so that they may be got at and be in condition to be planted with Lettuce by the end of February, or the first of March. By this time the weather is always mild enough to allow the sashes to be taken off from the Cabbage and Lettuce plants, and they are now transferred to the spare frames to cover and forward the Lettuce. Under each sash we plant fifty Lettuce plants, having the ground first well enriched by digging in about three inches of well rotted manure. The management of the Lettuce for heading is in all respects similar to that used in preserving the plants in winter; the only thing to be attended to, being to give abundance of air, and on the occasion of rain to remove the sashes entirely, so that the ground may receive a good soaking, which will tend to promote a more rapid and luxuriant growth.

The crop is fit for market in about six weeks from time of planting, which is always two or three weeks sooner than that from the open ground. The average price for all planted is about $4 per hundred at wholesale, so that again, with little trouble, our crop gives us $2 per sash in six weeks.

I believe this second use of the sash is not practiced outside of this district, most gardeners having the opinion that the winter plants of Cauliflower, Cabbage, or Lettuce, would be injured by their complete exposure to the weather at as early a date as the first of March. In fact, here we have still a few old fogies among us, whose timidity or obstinacy in this matter prevents them from making this use of their sashes, which thereby causes them an annual loss of $2 per sash, and as some of them have over a thousand sashes, the loss is of some magnitude.

In my own practice, I have made my sashes do double duty in this way for fifteen years; the number when I first started being fifty, increasing to the present time, when I have in use fifteen hundred sashes. Yet in all that time I have only once got my plants (so exposed) injured, and then only a limited number, which I had neglected to sufficiently harden by airing.

We have still another use of the sashes to detail. Our Lettuce being cut out by middle of May, we then plant five or six seeds of the Improved White Spine Cucumber, in the centre of each sash. At that season they come up at once, protected by the covering at night. The sashes are left on until the middle of June, when the crop begins to be sold. The management of the Cucumber crop, as regards airing, is hardly different from that of the Lettuce, except in its early stage of growth it requires to be kept warmer; being a tropical plant, it is very impatient of being chilled, but in warm days airing should never be neglected, as the concentration of the sun's rays on the glass would raise the temperature to an extent to injure, if not entirely destroy the crop. This third use of the sashes I have never yet made so profitable as the second, although always sufficiently so to make it well worth the labor.

There are a few men here who make a profitable business from the use of sashes only, having no ground except that occupied by the frames. In this way the winter crop of Cauliflower or Cabbage plants is sold at an average of $3 per sash, in March or April; the Lettuce at $2 per sash in May, and the Cucumbers at $1 per sash in June, making an average of $6 per sash for the season; and it must be remembered that these are wholesale prices, and that too, in the market of New York, where there is great competition. There is no doubt, that in hundreds of cities and towns of the Union, the same use of sashes would double or treble these results.

Cold frames are also used for sowing the seeds of Cabbage, Cauliflower, and Lettuce, instead of hot-beds; if the frames are closely shut up and covered at night by mats, the plants will be but little later than those from the hot-beds, and are raised with far less trouble. In sections of the country where these plants cannot be set out before May, it is useless to raise them in hot-beds. On the other hand, in the Southern States, where in the months of February and March there are no frosts, by adopting the same care in covering up at night, the seeds of Tomatoes, Peppers, and Egg plants, and the sprouts from Sweet Potatoes, can be forwarded with much less trouble in the cold frames than in the hot-bed.

CHAPTER IX.

FORMATION AND MANAGEMENT OF HOT-BEDS.

Although we do not consider hot-beds so convenient or even economical in the long run as the forcing houses, elsewhere described, yet, as beginners in the business are usually not over-supplied with means, and as hot-beds are to be had at much less first cost than the forcing houses, we give a description of their formation and management. The most convenient sash for the hot-bed is the 3×6 foot sash, made out of 1½ inch pine, costing here, at present prices, when painted and glazed, about $4 each. This is almost double the cost of what they were before the reign of high prices; but as we get corresponding rates for the commodities raised under them, we must not complain.

The frame for the hot-bed is usually made movable, in lengths which three sashes will cover, making, when complete, a box-like structure, 9 feet long, (the width of 3 ashes, 3 feet wide), and 6 feet wide, (the length of the sashes); at the bottom or lower part, the plank should be 15 or 18 inches high; the back or top, 24 inches; so that, when the sashes are placed on, it will give them the necessary angle to receive the sun's rays and throw off the rain. The sashes should be made as tight fitting as they will easily work, and the plank, forming the sides of the box, should be high enough to cover the thickness of the sash, n order to prevent the cold air from penetrating. This is one style of hot-bed frame, and the one most commonly used in private gardens; but in our market gardens, where a large surface is used, our necessities compel us to adopt a far more economical mode, both in the cost of the frame work and heating material. This is done somewhat after the manner adopted for Cold Frames. Parallel excavations are made, usually in lengths of 60 feet, 2½ feet deep, and 6 feet wide; the sides of these pits are boarded up with any rough boarding, nailed to posts, and raised above the surface 18 inches at the back, and 12 inches at front. Strips are stretched across, on which the sashes rest, wide enough to receive the edges of the two sashes where they meet, and allow of a piece of about an inch between them, so that the sash can be shoved backward and forward, and be kept in place in giving air, etc.

The heating material is next in order; this should be horse dung, fresh from the stables, added to which, when accessible, about one-half its bulk of leaves from the woods. The manure and leaves should be well mixed and trodden down in successive layers, forming a conical heap, large enough to generate fermentation in severe winter weather. Care must be taken that the material is not allowed to lie scattered and get frozen, else great delay will ensue before heat can be generated. A few days after the pile has been thrown together, and a lively fermentation has taken place, which will be indicated by the escape of steam from the heap, it should be again turned over and carefully shaken out, formed again into a pile, and left until the second fermentation occurs, which will be usually in two or three days. It may now be placed in the pit, being regularly beaten down by the back of the fork, and trodden so that it is uniformly of the same solidity, and to the required depth, 2½ feet. The sashes are now placed on the frames, and kept close until the heat rises; at this time a thermometer, plunged in the heating material, should indicate about 100 degrees, but this is too hot for almost any vegetable growth, and besides the rank steam given out by the fermentation, should be allowed to escape before operations of sowing or planting begin. New beginners are very apt to be impatient in the matter of hot-beds, and often lose the first crop by planting or sowing before the violent heat has subsided, which it generally will do in about three days, if the heating material has been sufficiently prepared. As soon as the thermometer in the frame recedes to 90, soil should be placed on, to the depth of 6 or 8 inches. This soil must be previously prepared, of one-third well rotted manure, (or, if procurable, rotted refuse hops, from breweries), and two-thirds good loam, spread regularly over the surface of the hot-bed.

We use hot-beds for various purposes. One of the most important uses is the forcing of Lettuce; this is planted in the hot-beds, (from plants grown in the cold frames), 50 under each sash, the first crop by 2nd week in January; it is covered at night by straw mats, and is usually marketable by the first of March. At that season Lettuce is always scarce, and will average, if properly grown,

$8 per 100, or $4 per sash. The crop is cut out by the first week in March, giving plenty time to plant the same hot-bed again with Lettuce; but now it is no longer a hot-bed, for by this time the heat from the dung is exhausted, and it is treated exactly as described in the chapter on Cold Frames.

Another use for the hot-bed is the raising of Tomato, and Egg, and Pepper plants. The bed should be prepared for these, not sooner than the 2nd week in March, and of temperature, about the same as before described. In sowing, it is well to cover the seed with some very light mold; nothing is better than leaf mold and sand, patting it gently with the back of the spade From the time

Fig. 21.—WATERING POT.

the seed is sown, attention to airing, during the hot part of the day, and covering up at night, is essential, and also that the soil be never allowed to get dry. The watering should be done with a very fine rose Watering Pot, (fig. 21), and with tepid water. The temperature at night may range from 55 to 65°, and during the day from 70 to 80°. As soon as the seedling plants are an inch or two high, which will be in 5 or 6 weeks, they must be taken up and re-planted in a more extensive hot-bed, for they now require room. Tomatoes should be planted of a width to give 75 or 100 in each sash. Pepper and Egg Plants do better if planted in small flower pots, (3-inch), as they are more difficult to transplant; they may now also be kept a little closer in the hot-bed than the Tomatoes, as they require more heat. After transplanting, great care is necessary that they always be immediately watered, and shaded from the sun until they have struck root, which will be in 2 or 3 days after transplanting.

The hot-bed is also the medium for procuring us Cabbage, Cauliflower, and Lettuce plants, for early outside planting, when not convenient to winter them over as described in the uses of cold frames. The seeds of these are sown about the last week in February, are treated in all respects, as regards covering up at night, as the Tomatoes, etc.; but being plants of greater hardiness, require more air during the day. They will be fit to plant in the open garden by the middle of April. The beds they are taken from are usually employed for the re-planting Tomatoes, which it is not safe to plant, in the open ground here, before the middle of May.

Sweet Potato plants are almost universally raised in hot-beds, but as this is a plant that luxuriates in a high temperature, the hot-bed should not be formed to start them until the middle of April. The soil should be a mixture of sand and leaf mold, laid on of the usual thickness on the hot-bed, 6 inches. The tubers are placed closely together, and the same sandy compost sifted over them to the depth of two inches; some split the large ones lengthwise, and place them flat side down; they should not be watered until they start to grow. They are fit to plant out about six weeks after starting.

Two most essential points in working hot-beds are, in

covering up at night, and in giving air during the day. It often happens that a few mild nights in March or April delude us into the belief that all the cold weather is over, and the covering up is in consequence either carelessly performed or abandoned. Every season shows us scores of victims to this mistake, who, by one cold night, lose the whole labors of the season. It is always safest to cover up tender plants, such as Tomatoes, Sweet Potatoes, etc., until the 10th of May in this latitude, and the more hardy plants, such as Cabbage, to the 1st of April, when raised in hot-beds; even if there is no danger from freezing, it will give a more uniform temperature, and consequently conduce to a more healthy growth. The want of close attention in airing is equally dangerous; often an hour's delay in raising the sashes, will have the effect of scorching up the whole contents of the hot-bed, and irregularity of airing will always produce "drawn" and spindling plants, even when they are not entirely killed. The thermometer is the only safe guide, and should be regularly consulted, and whenever it indicates 75°, it is safe to admit less or more of the outer air, proportioned of course to the condition of the atmosphere; if there be bright sunshine, and cold wind, very little will suffice; if calm, mild, and sunny, admit larger quantities.

Coverings for Protection against Frost.—To cover up hot-bed sashes, we use either light pine shutters or straw mats; the shutters are made the exact size of the sash; there is no necessity of their being more than half an inch in thickness, as that is quite as effective in keeping out the cold as two inches would be, and they are much cheaper and more convenient to handle. Straw Mats are, however, by far the warmest covering, and in hot-bed culture are almost indispensable. They are always made at home, during wet days or stormy weather in winter. The manner of making them is very simple, and will readily be learned at the first attempt. The "uprights," (or warps), are formed of five strands of a tarred string, known as "marline;" these are tightly strained 10 inches apart, by being attached to five strong nails at bottom of a wall, corresponding with the same number 7 feet from the bottom. Against these strings (beginning at the bottom) are laid small handfuls of rye straw, the cut side

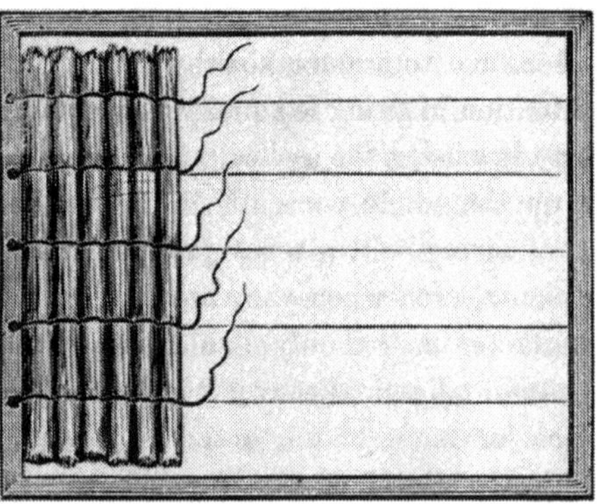

Fig. 22.—MAKING A STRAW MAT.

out, as long and straight as can be procured; this is secured to the uprights by a lighter kind of tarred string, by taking a single turn around the upright and the straw, and so continued until the mat is finished. Some use a frame to which the strings, forming the warp, are attached, as shown in fig. 22. This allows the operator to have his work upright or horizontal, as may be most convenient. Two workmen will make about five mats in a day.

When finished, the mats should be 7 feet in length and 4½ feet in width, two being sufficient to cover three sashes. The reason for having them made one foot longer than the sash is, that there may be 6 inches to overlap at top and bottom, which are the most necessary points to secure from frost. In making these mats they may be constructed of sedge from the marshes, or salt meadow hay, when rye straw cannot be procured. It is important, however, that they may be made as light as possible, one inch in thickness being quite sufficient. By care in handling them, these mats will last for six or eight years.

CHAPTER X.

FORCING PITS OR GREEN-HOUSES.

Forcing pits and green-houses of the style about to be described, whenever the greater expense in their erection is not a consideration, are, in our experience, far superior, and in the course of five or six years, more economical for all purposes of forcing or forwarding vegetables, than the hot-bed or cold frame. Figure 23 represents the end section and ground plan of the style we have in use, and which may be adapted to any plant that requires artificial heat and protection of glass. The pits, represented by this plan, are 100 feet in length, and each 11 feet wide inside. The heating is accomplished by one of Hitching's Patent Boilers, (*C*), heating about 1200 feet of 4-inch pipe. The glass roof, (*E*), is formed of portable sashes, 6 × 3 feet; each alternate sash is screwed down, the others being movable, so as to give abundance of air. The fixed roof plan of building green-houses or pits, is, in our climate, a great mistake, in my opinion, unless in large graperies or conservatories, where architectural beauty is of more consideration than the health of the plants. The mov-

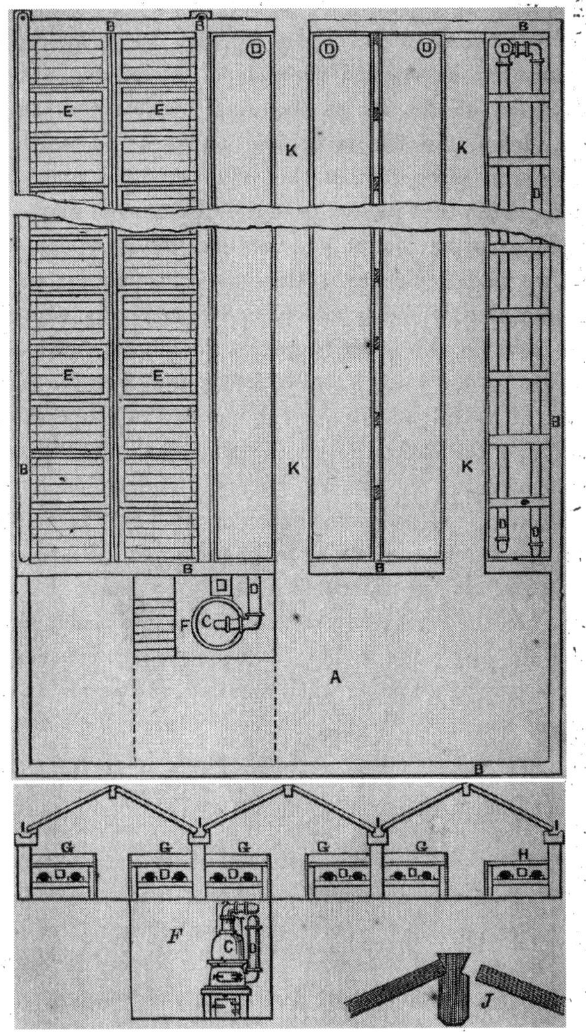

Fig. 23. — END SECTION AND GROUND PLAN OF FORCING PITS HEATED BY WATER PIPES.

able sash is elevated, to admit air, by an iron bar, 15 inches long, attached to the sash by a staple; into this bar is punched three holes, so to regulate the admission of the air as required. When the sash is shut down, the bar is hooked on to a pin which secures it in place, so that the sash cannot be moved by wind. I am thus particular in describing what may seem a simple matter; but this system of airing is not common, and we made some stupid blunders before we hit on our present plan, which is cheap, simple, and very effective. The movable sash is not hinged at the bottom, but is held in place by two small plates of iron screwed on the gutter plate. The ridge-poles are cut out of the shape shown at *I*, and the sashes lie on the shoulder. The interior arrangement of the pits will be easily understood by the end section. *G*, shows the bench or table as it is completed; this shows the boxing-in of the pipes, (*D*), to give "bottom heat" to the cuttings, seeds, or plants, that are placed on the bench, but on the sides of the bench, along the walk, one plank is hinged throughout the house, so that it may be let down when required, and permit the escape of heat into the pit. The walks through the house, (*K*), are 2 feet wide. A brick shed, (*A*), covers the boiler pit, (*F*), and is attached to the north end of the pits. Besides breaking the wind at this vulnerable point, we find this shed a most excellent place for many purposes, as it is kept from freezing by the heat that escapes from the boiler pit, which would otherwise be lost. This heat may be rendered to a very profitable account in forcing Mushrooms or Rhubarb, if desired for that purpose.

The system of attaching three pits together, if not new

in this country, is certainly new in its almost universal adoption by commercial gardeners, in all houses erected during the past five years in the neighborhood of New York; it has great advantages over the detached system; being less expensive in heating, more saving of space, and, above all, far more economical in cost of construction. I prefer having only three together, for the reason that, when we have the snow to clear away, it is quickly done by being shoveled from the two valleys or furrows over the ridges; although we have one grower in this neighborhood who has 12 connected houses, and finds but little trouble with snow; our snows being mostly from the North, the shed breaks them off in a great measure, and what blows over, blows mostly off through the valley between the sashes. The water from the gutter is led into a cistern, at the south end of the green-houses, of a capacity of not less than 3000 cubic feet, if 5000, all the better; to this is connected a West's Force Pump, fig. 24, with 150 feet of 1½-inch hose, and to the end of the hose is attached a heavy sprinkler. One man pumps, and another regulates the water and sprinkles it over the plants. My estab-

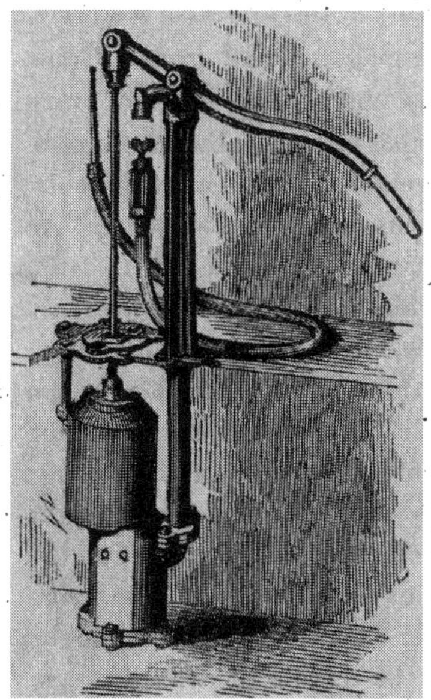

Fig. 24.—WEST'S FORCE PUMP.

lishment contains over an acre of glass, and yet, by this labor-saving arrangement, all the plants are thoroughly drenched with water by two men in four hours. Before adopting this method, which I only did last year for the first time, four hands were employed the whole day during the spring months in watering, and then the work was not done half so well. There is nothing that I have ever done connected with horticultural operations, that has been so entirely satisfactory as this system of watering.

In these pits may be propagated and grown Grape Vines, Roses, and green-house and bedding plants of every description, in the best possible manner. But as our present purpose is only with vegetables, I will endeavor to describe our mode of operations with some of these. As Lettuce, from the great quantities consumed in all large cities, is now, and will be likely to be, one of the most profitable vegetables to force, we begin, for our first crop, by sowing the seed about the first of September, in the open ground, of the Tennis Ball, or Black-seeded Butter sorts. These are planted on the benches of the forcing house in five or six weeks after, at about six or eight inches apart each way, on well enriched soil, placed on the benches to the depth of five or six inches. At this season, no "forcing" is required, in fact, if the sashes could be taken completely off until the middle of November, so much the better, but when it is not convenient, the sashes should be kept raised to admit air, night and day, until frosts begin to be severe; then they should be shut up at night, but no fire heat should be applied until the weather has been severe enough to indicate 32 or 34 inside the pits, and even then very slight, for if thev can be brought to maturity at this season without the temperature exceeding forty at night, (by fire heat), the crop will be all the better. The great thing in forcing all plants of this hardy nature being to avoid a *high temperature.* This first crop is usually ready by middle of December, and is cut off and sold in two weeks; the soil on the benches is slightly manured, dug up, and again planted (from plants sown in cold frames, or in boxes in the same pit) about November 1st. This second, or winter crop, requires more attention in growing, both in firing, watering, and airing, as it matures about March 1st, and consequently has had to be cared for during the coldest part of the year. The third crop, treated exactly as the second, is planted soon as the other is cut off, and matures about May 1st. We now vary the use of the pit, by planting at distances of about three feet apart along the centre of the bench, plants of the White Spine Cucumber, from seed sown about April first, in a corner of the pit that has been kept closer and warmer than that for the Lettuce; these are planted in pots about 3 inches in diameter, and by the time the benches are cleared in May, are fine strong plants, that gives a full crop during the month of June—fully a month sooner than from the open ground. The combined value of these four crops will average about $500, for an erection 100 feet in length by 11 feet in width. The estimated expense of cultivation is:—

Interest on $1000, cost of construction, at 10 per cent......	$100
Coal, 5 tons..	50
Labor, Manure, etc..	100
	$250
Receipts..	500
Nett Profit...	$250

These forcing pits are likewise used for starting seeds of Tomato, Egg, Pepper, Cabbage, and Cauliflower, and sprouting Sweet Potatoes, which is done with far less risk and in a much better manner, than can be accomplished by the hot-bed. One great advantage is, that by being able to walk inside of them, these pits are accessible in all weathers, while with the hot-bed or frames, we are in winter often debarred from examination for whole days together.

At present prices, in this locality, these pits cost about $10 per lineal foot, everything complete put up in the way indicated by the plan in a plain substantial manner. But those whose circumstances do not admit of the expense of heating by hot water, (which is nearly half the cost of the whole), may put up erections of exactly the same character, and heat them by the common smoke flue, at an expense of from five to six dollars per lineal foot, in the manner shown by the plan, fig. 25. It will be seen by this sketch, that two flues only are used for the three pits, each passing first up under the bench on the outside houses, is carried along the end and returned through the middle houses; this equalizes the temperature in all three, for the outside houses get only one run of the flue, but it being directly from the fire, gives about the same heat to the outside houses as two runs in the middle house, which being at a greater distance from the fire, are much colder. Three attached houses, heated thus, should not be over 50 feet long, in this latitude. Southward they may be 60 feet, and northward 40 feet. Peculiarities of locality have much to do with the heating; in positions particularly

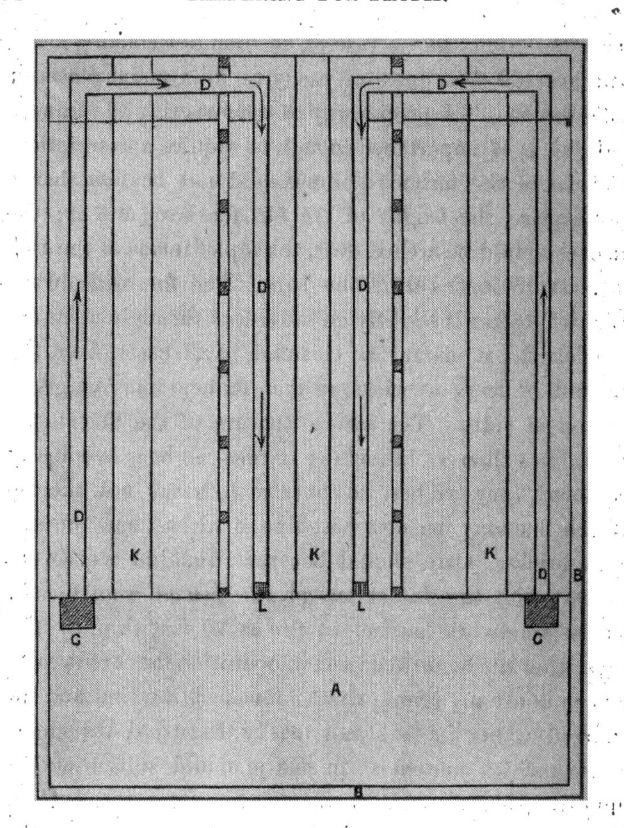

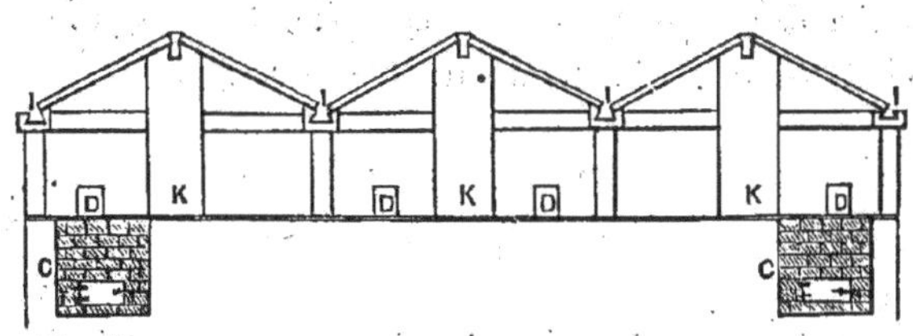

Fig. 25.—END SECTION AND GROUND PLAN OF FORCING PITS HEATED BY FLUE.

sheltered from the northwest, the same amount of flue will heat 60 feet, quite as easily as in exposed places it will heat 40. The proper way of constructing the furnace and flue, is of importance enough to require a description. The size of the furnace doors should not be less than 1 foot square, the length of the furnace bars, 2 feet; the furnace should be arched over, the top of inside of the arch at least 18 inches from the bars. The flue will always "draw" better if slightly on the ascent throughout its entire length; it should be elevated, in all cases, from the ground, on flags or bricks, so that its heat may be given out on all sides. The inside measure of the flue should not be less than 8×14 inches; if tiles can be conveniently procured, they are best to cover with, but if not, the top of the flue may be contracted to 6 inches, and covered with bricks. Care should be taken that no woodwork connect with the flue at any place. I have known cases where woodwork has caught fire at 70 feet from the furnace, after the house had been in operation for three years; but an unusually strong draft intensified the heat,and the charred timber ignited and totally destroyed the green-house and its contents. In the plan and section on the opposite page, *A*, is the shed, enclosing the furnaces *C C*; from which pass the flues, *D*, in the direction shown by the arrows to the chimnies, *L*. The benches are not shown here, but they are arranged as in fig. 23.

CHAPTER XI.

SEEDS AND SEED RAISING.

If there is one thing of paramount importance in vegetable gardening, it is purity of seed; and for this reason every seed that it is practicable for us to raise for our own use, we grow, no matter what the cost may be. On one occasion, our indispensable Wakefield Cabbage seed failed, from some peculiarity in the season, and there was no alternative but to buy from the seed stores; every store in New York was tried, but not a seed proved to be of the Wakefield, as we know it. One old gentleman, who always provided for such emergencies, had a two-year old reserve supply. I offered him $50 per pound, but could not procure an ounce from him. He too well understood the state of the case, and planted his whole ground with this variety, and as he got in ahead of all by nearly ten days, made a little fortune by the operation. That was about ten years ago; but I have never known a Jersey gardener to be out of this variety of seed since, and not know where to get it. On another occasion one pound of seed, purchased as Silesia Lettuce, and planted in my forcing frames, proved to be the curled

India Lettuce, useless, except for our hottest summer weather, and perfectly worthless for forcing. This was the most serious loss from bad seeds I ever encountered, amounting to at least $1000. Last year, quite a number of my neighbors lost heavily in purchasing seed of the *erect* variety of Thyme, instead of the *spreading* variety; the crop being all but worthless in consequence. No wonder then that the market gardeners are so skeptical about using seeds except those of their own raising, or from those of their immediate neighbors, in whose knowledge and honesty they have entire confidence.

There is but little new to say of the manner of raising seeds; the importance of selecting the purest specimens of each variety, and of keeping plants that are of the same families, as far distant apart as the limits of the ground will admit, is now well understood. It is not practicable, however, to raise all the seeds wanted in our vegetable gardens, in our climate, and consequently we have to rely on importation for seeds of Brocoli, Cauliflower, some varieties of Cabbage, Radishes, Peas, etc. But the great majority of seeds used are raised here, our climate being particularly well adapted for maturing them. In the raising of market vegetables, near large cities, the usual practice is, for each grower to grow only a few varieties, and these of the sorts most profitable to his location or soil. For example, we of New Jersey, in the immediate vicinity of New York, grow Beets, Cabbages, Cauliflower, Cucumbers, Lettuce, Radishes, and Turnips, as a first crop; followed by Celery, Thyme, Sage, Brocoli, and late Cabbage, as a second crop. Of these varieties we save all the seeds that it is practicable for us to raise; while the more Southern counties of the State, where crops mature ten or twelve days earlier, but the distance greater from market, the bulkier and cheaper articles are not grown, and only the more portable and (when early) valuable kinds are raised, of which Tomatoes, Melons, Peas, Kidney Beans, Early Turnips, and Beets, are the staple articles. There, also, the growers know well the necessity of sowing only such seeds as are grown by themselves, or from sources that they know to be reliable.

Seed growing, as practised by market gardeners, is on much too small a scale to make it profitable; in fact, there is hardly a seed we raise, but costs us much more than what we could purchase it for from the seedsmen. Seedsmen are supplied by regular seed growers, who make a special business of it; they are located principally in the Eastern States, and devote many thousands of acres of the finest lands to the purpose. They are a highly responsible class of men, who thoroughly understand the business, and are now successfully competing with the English and French growers, from whom, only a few years ago, nearly all our seeds were imported. Just so soon as our seedsmen are able to get their entire supply from reliable men here, there will be no necessity for the market gardeners continuing to be their own seed growers; they would also greatly conduce to the increase of their business by taking the trouble to ascertain the varieties most suitable for market purposes. Above all, no seed should ever be sold without its germinating qualities being thoroughly tested. Neither should any gardener risk his crop without testing the seed, unless he has implicit confidence in the source from whence it has been purchased.

It will be understood, that of all annual plants, such as Beans, Corn, Cucumbers, Egg Plants, Lettuce, Melons, Peas, Radishes, Tomatoes, etc., the seed is saved the season of planting, and should be always taken from those first maturing, if earliness is an object. The seeds of biennial vegetables, such as Beets, Carrots, Celery, Cabbages, Onions, Leeks, Parsley, Parsnips, etc., are raised by selecting the best specimens from those preserved over winter, planting them out in good soil on the opening of spring, at distances such as are recommended for their growing.

Duration of Germination in Seeds.—There are very few seeds that will not germinate as freely the second year as the first, if properly kept in a cool place, and not exposed to either a too drying or too damp an atmosphere. With the exception of Parsnips, Onions, and Leeks, I would just as confidently sow seed two years old, as when fresh gathered; but there is a limit to the vitality of seeds, varying much in the different species.

Among those, only safe for *two* years, are: Beans and Peas, of all kinds; Peppers, Carrot, Egg Plant, Okra, Salsify, Thyme, Sage, and Rhubarb.

Those safe for *three* years: Asparagus, Endive, Lettuce, Parsley, Spinach, and Radish.

Those safe for *four* years: Broccoli, Cauliflower, Cabbage, Celery, and Turnip.

Those possessing the greatest vitality are: Beet, Cucumber, Melon, Pumpkin, Squash, and Tomato; the time ranging from five to ten years.

We often find this knowledge very valuable; for example, in procuring the stock of a seed *said* to be good, of a variety that does not seed the season it is marketable, such as Broccoli, Cauliflower, Cabbage, or Celery, we procure enough to last at least two seasons; the first season only a little is sown, to test the merit of the variety, (for we are never incautious enough to risk a full crop with one experiment); if it proves valuable, we have enough in reserve to sow for a full crop, knowing that it is sure to germinate. This was particularly the case with our New Dwarf Celery; on the recommendation of a friend I imported ten pounds of the seed, but doubtful of how it would suit our market, only as much was sown as would furnish a few hundred plants. These showed so much superiority, in all respects, to the tall varieties that we had been growing, that the following season I put in half my crop with the dwarf seed. The thing was entirely new in our market, and so much superior, that it sold for prices that would seem incredible. My ten lb. bag was not half exhausted, and the next season I planted my whole crop, fourteen acres, containing nearly half a million roots, and made one of the best hits I ever made in gardening. But by this time my neighbors began to take an unusual interest in my Celery crop, and I could monopolize the variety no longer. New Yorkers will this season appreciate Celery more than ever before, and in consequence far more will be sold, for there is as marked a difference in the flavor of this variety and the coarse growing kinds, as between a Bartlett and a Choke pear.

CHAPTER XII.

HOW, WHEN, AND WHERE TO SOW SEEDS.

As seed sowing is the starting point of cropping, a thorough knowledge of the conditions necessary for the germination of the different varieties, will go far towards putting the tyro in gardening well on the way to success. The very general want of knowledge in this matter is too often the cause of much undeserved censure upon the seedsman, for in nine cases out of ten the failure is not with the seeds, but results from the *time* or manner of planting. When the owner of a garden sends his order for seeds to the seedsman, it is generally a complete list of all he wants for the season. They are received, and the interesting operation of sowing is begun: first in a hot bed, if he has one, often as early as the first week in February, (a month too soon by the way), and in go indiscriminately, at the same date, and under the same sash his seeds of Cabbage, Cauliflower, Lettuce, and Egg Plant, Peppers and Tomatoes. Yet even in the waning heat of this early hot-bed, where a thermometer would possibly not indicate more than fifty degrees, he finds in a week or so his Cabbage, Lettuce, and Cauliflower "coming through" nicely, but as yet no Egg Plants, Peppers, or Tomatoes. He impatiently waits another week, makes an examination, and discovers that instead of his Tomatoes and Egg Plants beginning to vegetate, they are commencing to rot. It is now plain to him that he has been cheated; he has been sold old seed, and if he does nothing worse, he for ever after looks upon the seedsman he has patronized as a venial wretch, destitute of principle and honesty. But he must have Tomatoes, Peppers, and Egg Plants, and he buys again, from another seedsman, warranted honest. He renews his hot-bed, it is now a month later, and a bright March sun, with milder nights, gives him the proper temperature in his hot-bed—70 or 80°, and his eyes are at last gladdened by the sprouting of the troublesome seed. April comes with warm sunshine, inviting him to begin to "make garden" outside. He has yet the balance of the original lot of seeds that he bought in February. But as he is still entirely befogged about the cause of his failure in the first hot-bed, he begins his open ground operations with little confidence in his seeds, but as he has got them, they may as well be tried. And again he sows in the same day his Peas and Lima Beans, Radishes and Pumpkins, Onions and Sweet Corn. Hardy and tender get the same treatment. The result must of necessity be the same as it was in the hot-bed the hardy seeds duly vegetate, while the tender are of course rotted. This time he is not surprised, for he is already convinced that seedsman No. 1 is a rascal, and only wonders how any of his seeds grew at all, so he again orders from seedsman No. 2 for the articles that have failed. Here circumstances continue to favor the latter, for by this time the season has advanced in its temperature and the seeds duly vegetate. Every farmer knows that, in this latitude, he can sow Oats or Wheat in March or April, but that if he sows his Corn or Pumpkins at the same time, they will perish; this he knows, but he may not know that what is true of the crops of the farm, is equally true of the garden. Hence the importance of a knowledge of the season when to sow vegetable seeds, or set out plants.

The temperature best fitted for the germination of seeds of the leading kinds, will be best understood by the tabular form given below.

Vegetable seeds that may be sown, in this latitude, from the middle of March to the end of April. Thermometer in the shade averaging 45 degrees.		*Vegetable seeds that may be sown in the open ground, in this latitude, from the middle of May to the middle of June. Thermometer in the shade averaging 60 degrees.*	
Beet.	Lettuce.	Lima Beans.	Water Melon.
Carrot.	Parsley.	Bush Beans.	Squash.
Cress.	Parsnip.	Cranberrry	Pumpkin.
Celery.	Onions.	Pole Beans.	Tomato.
Cabbage.	Peas.	Scarlet Runner Beans.	Nasturtium.
Cauliflower.	Radish.	Sweet Corn.	Okra.
Endive.	Turnip.	Musk Melon.	Cucumber.
Kale	Spinach.		

It will be understood that these dates refer only to the latitude of New York, farther South, operations should be begun earlier,—farther North, later. So much for the time of sowing; I will now refer to suitable soil and the manner of sowing.

The Choice of Soil, when choice can be made, is of great importance, the best being a light soil, composed of leaf mold, sand, and loam; the next substitute for leaf mold being well decayed stable manure, or better yet, decayed refuse hops from the breweries, in short, anything of this nature that will tend to lighten the soil, the point to be avoided being a *weight* of soil, either from the nature or quantity of it. The nature of the soil is not of so much importance for the germinating of large vigorous seeds, as Peas, Beets, Beans, Corn, etc. But with the delicate, slow sprouting sorts, as Celery. Parsnip, Egg Plant, or Peppers, it is of much importance Seeds of nearly every garden vegetable should be sown in rows; the distance apart, according to the variety, and the depth proportioned to the size of the seed. No better information can be given in this matter, than the old rule of covering the seed with about its thickness of soil, but this should always be followed up by having the soil pressed closely down. In our market gardens here, we invariably have the ground rolled after sowing, or in frames or hot-beds, where the roller cannot be used, we pat the soil evenly down with a spade after sowing. This may not be of so much consequence in early spring, when the atmosphere is moist, but as the season advances, it is of great importance. I have seen many acres of Carrots and Parsnips lost for want of this simple attention; the covering of the seeds being loose, the heated air penetrates through, drying the seeds to shriveling, so that they never can vegetate. My farmer readers, no doubt, have had plenty of similar experiences with Turnips, where they have been sown broadcast without rolling. Another advantage in rolling after seed sowing is, that it leaves the surface smooth and level, thereby lessening greatly the labor of hoeing. Instead of adopting the questionable practice of steeping seeds, preparatory to sowing in dry hot weather, we prefer first to thoroughly saturate the bed with water, and after it has dried enough, so that it can be raked without clogging, sow the seed. It is much better to do this than water after the seed has been sown, as it has a tendency in most soils to bake the surface.

Sowing in Hot-beds.—It would lengthen this chapter too much to give extended directions for sowing seeds in hot-beds. I will briefly say, that after the hot-bed has been formed—say by the first week in March, let soil, of the kind recommended, be placed on it six inches deep, into which plunge a thermometer three or four inches, and when the temperature *recedes* to 75 degrees or 80 degrees, you may then sow, giving air in mild weather as soon as the seeds begin to vegetate, covering up warmly at night by mats, straw, or hay. But many that may read this never saw a hot-bed, and are perhaps never likely to have one: to such I would say that there is an excellent substitute on hand in most dwellings, in the kitchen or basement windows, facing South or East, inside of which is a temperature usually not far from that required for the vegetation of seeds, and where plants from seeds of the early vegetables, or tender plants for the flower border, may be raised nearly as well, and with far less attention, than in a hot-bed. Instead of hot-beds, we use our green-houses for the purpose, using shallow boxes in which we sow the seed; these are made from the common soap box, cut in three pieces, the top and bottom forming two, and the middle piece, bottomed, making the third; these form cheap, convenient boxes. Fill these nearly full with the soil recommended, and after sowing, press nicely down level, and make the surface soil moderately firm; keep moist, in a temperature in the window of from 60 to 70 degrees, and your little trouble will soon be rewarded.

In this way seeds should be sown thickly, and after they have made the first rough leaf, should be again planted out into the same kind of box, from one to four inches apart, according to the kind, and placed in the window to receive similar treatment as the seeds; but as the season advances, in mild days they should be set out of doors, care being taken that they are brought in at night, and that the soil in the boxes is never allowed to get dry.

I know what is usually the first thing the novice in gardening does if he gets any choice seed or favorite cutting; he has some how got the belief that there is some hidden virtue in a flower pot, and he accordingly sows his seed or plants his cutting in a pot, but in nine cases out of ten they are destroyed, or partially so, by the continued drying of the soil in the porous flower pot. If early in the season, let delicate seeds be sown in the kitchen or sitting room window, in the boxes as recommended, or if late, in the open border; but seeds should never be sown in pots, as even in experienced hands they are much more troublesome and uncertain than boxes.

Rotation of Crops.—Cultivators of the most limited experience soon discover that the same kind of crop cannot be grown on the same ground for many years in succession, without deterioration. A great many theories have been advanced assigning causes for this, but they are as yet far from satisfactory, and as this is not a book of theory but of practice, I will not further allude to them. The following general rules will be found useful as a guide:—

1st. Plants, of the same family, should not be planted to succeed each other.

2d. Plants, which occupy the ground for a number of years, such as Rhubarb and Asparagus, should be succeeded by annual plants.

3d. Crops, grown for heads, such as Cauliflower, Cabbage, etc., should be succeeded by crops grown for their bulbs or roots.

It is hardly practicable to vary crops according to any set rule, the demand in different localities for certain articles being greater than in others. Generally variety enough is demanded to allow of sufficient rotation. Our system of heavy manuring, deep culture, and taking two crops from the ground each season, seems to do away in a great measure with the necessity for systematic rotation, which would often be found to be very inconvenient. The crops of all others, that we find most benefited by change are, the Cabbage tribe, together with the allied Turnip, Radish, etc., while on the other hand we have grown Onions successively on the same ground for ten years—the last crop just as good as the first; but as a rule of safety, vary whenever you can.

Quantity of Seeds per Acre.—The quantities given below are somewhat higher in some kinds than the usual estimates, our experience showing us that in weak vegetating seeds, such as Parsnips, Carrots, etc., it requires numbers, particularly on stiff soils, to force through plants enough to form a crop; more seed is required when sown during the dry hot months of summer, than if sown in the cooler and moist seasons of spring and fall, hence quantities are regulated accordingly:—

QUANTITY OF GARDEN SEEDS PER ACRE.

		About.
Beans, Dwarf Kidney	in drills	1½ bushels
Beans, Pole	in hills	12 quarts.
Beets	in drills...8 pounds or 8	quarts.
Carrot	in drills	5 pounds.
Corn, (shelled)	in hills	2 quarts.
Cucumbers	in hills	1 pound.
Melon, (Musk)	in hills	1 pound
Melon, (Water)	in hills	1½ pounds.
Onions, (for bulbs)	in drills	6 pounds.
Onions, (for sets)	in drills	30 pounds.
Onion sets. (small)	in drills	10 bushels.
Potatoes, (cut tubers)	in drills	10 bushels
Parsnips	in drills	8 pounds.
Peas	in drills	1½ bushels.
Radish	in drills	5 pounds.
Radish	broadcast	10 pounds.
Spinach	in drills	10 pounds.
Salsify	in drills	10 pounds.
Squash	in hills	1 pound.
Turnip	in drills	2 pounds.
Turnip	broadcast	4 pounds.

QUANTITY OF SEEDS REQUIRED FOR A GIVEN NUMBER OF PLANTS.

	About.		*About.*
Asparagus, 1 oz	500 plants	Pepper, 1 oz	1000 plants
Cabbage, 1 oz	2000 "	Tomato, 1 oz	1500 "
Cauliflower, 1 oz	2000 "	Thyme, 1 oz	5000 "
Celery, 1 oz	3000 "	Sage, 1 oz	1500 "
Leek, 1 oz	1500 "	Savory, 1 oz	2000 "
Endive, 1 oz	3000 "	Marjoram, 1 oz	1500 "
Egg Plant, 1 oz	1000 "	Rhubarb, 1 oz	500 "
Lettuce, 1 oz	3000 "		

FARM SEEDS PER ACRE.

	About.
Wheat, broadcast	1½ bushels.
Barley, "	2 to 3 "
Oats, "	2 to 3 "

Buckwheat, broadcast	1 bushel.
Indian Corn, (for soiling)	3 "
Hemp	1½ "
Flax	1½ "
Peas	3 "
Vetches	3 "
Chinese Sugar Cane	12 quarts.
Broom Corn	10 "
White Clover, (alone)	15 pounds.
Red Clover, (alone)	20 "
Lucerne, (in drills)	15 "
Blue Grass, (alone)	3 bushels.
Rye Grass, (alone)	2 "
Orchard Grass, (alone)	3 "
Timothy Grass, (alone)	½ "
Red Top Grass	3 "
Mixed Lawn Grass	3 "
Clover, } together for one acre	10 pounds Clover.
Timothy, } together for one acre	½ bushel Timothy.
Red Top, } together for one acre	1 bushel Red Top.

THE NUMBER OF PLANTS, TREES, ETC., REQUIRED TO SET AN ACRE.

Distance	*Number.*	*Distance*	*Number.*
1 ft. by 1 ft.	43,560	6 ft. by 6 ft	1,210
1½ ft by 1½ ft.	19,360	9 ft by 9 ft.	537
2 ft. by 2 ft.	10,890	12 ft. by 12 ft.	302
2½ ft. by 2½ ft	6,970	15 ft. by 15 ft	194
3 ft. by 1 ft.	14,520	18 ft by 18 ft.	134
3 ft. by 2 ft	7,260	20 ft. by 20 ft.	103
3 ft. by 3 ft.	4,840	25 ft by 25 ft.	70
4 ft. by 4 ft.	2,722	30 ft. by 30 ft.	40
5 ft. by 5 ft.	1,742	40 ft. by 40 ft	27

CHAPTER XIII.

TRANSPLANTING.

TRANSPLANTING is an operation of great importance; the condition of the plant, the state of the soil, and of the atmosphere, have much to do with its success, independently of the simple mechanical operation. It is not very easy to instruct the uninitiated as to what the proper condition of the plant should be; experience in this being, as in everything else, the best teacher. Attention to keeping the seed-bed clear of weeds, the "topping" of plants when they get too tall, and careful digging up of them so as to preserve the root fibres, will all greatly assist. We cannot always get the soil in the proper condition of moisture to receive plants at the time transplanting should be performed, but to make up for the want of moisture, planting should be delayed always until late in the afternoon, unless in cloudy weather. It is also of great importance that the ground be freshly plowed; the moist soil thus brought to the surface will induce the formation of root fibres in one night, after which the plants are comparatively safe; but if they be allowed to wilt before the new roots begin to be emitted, and continued dry weather ensues, then nothing will save them but having recourse to watering, which should always be avoided if possible. In planting, each man is provided with a boy, who carries the plants in a basket, and whose duty is to drop the plant on the line at the proper distance before the planter. In planting, a hole is made by the dibber about the depth of the root; the plant being inserted, the soil is then pressed close to the root, the hole thus made by the displacement of the soil is again filled up by one stroke of the dibber. In dry weather we still further firm the plant, by each planter returning on his row, and treading the soil around plants firmly with the feet. I am thus particular in describing a simple matter, knowing well, that millions of plants are annually lost by inattention to this firming of the soil. The same rule is applicable to transplanting of all kinds, trees, shrubs, or vegetables. Instead of "puddling" the roots in mud, we prefer to keep the plants dripping wet during the whole time of planting, so that each plant, as it is put in the soil, puddles itself by the particles of soil adhering to the wet root. Besides, the leaves of the plant, being wet, will for some time resist the action of the dry air.

CHAPTER XIV.

PACKING OF VEGETABLES FOR SHIPPING.

This is a matter for which it is not very easy to give directions, as the distance, season, and articles to be packed will greatly determine the manner in which it is to be done; but a few general directions may assist the inexperienced.

The mode of packing during spring and summer is almost entirely the reverse of that practiced during autumn or winter, for the reason that, when the temperature is high, provision must be made in the package for the admission of air to prevent the articles from heating; while in cold weather, when there is but little danger from heating, but more to be apprehended from frost, close packages must be used accordingly.

As early vegetables are always shipped from a warm climate to a colder one, at a season which, of course, must be warm to mature them, open work baskets or slatted boxes must be used. If barrels are used, care must be taken that openings be made plentifully in the sides, so that air may be admitted. For distances requiring a delay of more than forty eight hours in the transit, for most articles, barrels are too large; boxes or baskets, one-fourth the capacity of a flour barrel, being safer. The articles shipped in this manner from southern ports to northern markets, are: Asparagus, Beans, Cucumbers, Lettuce, Melons, Peas, Radishes, Tomatoes, and other summer crops. Bulky articles, such as Cabbages, Beets, Sweet Corn, Water Melons, Turnips, are often shipped loose on the decks of steamers, sloops, etc.; but even then, care must be taken that the heaps are not too large, else they may be injured by heating. The judgment of the shipper must be exercised in respect to the article to be shipped. Articles that lay close, will require to be shipped in smaller packages than those that lie so loosely that the air can pass among them; for example, Melons may be safely packed in a barrel, while, if Tomatoes were so packed, they would be utterly destroyed.

The winter or fall shipping of vegetables is the reverse of the summer, for then we send from the North to the South, our colder and damper atmosphere being more congenial to the growth of late crops. Close packages are now used, but still not too large; barrels being best suited to such articles as Beets, Carrots, Celery, Onions, Parsnips, Potatoes, or Turnips, while Cabbages and Cauliflowers may be shipped in crates or in bulks.

CHAPTER XV.

PRESERVATION OF VEGETABLES IN WINTER.

Our manner of preserving vegetable roots in winter is, I think, peculiar to this district, and is very simple and effective.

After taking up such crops as Beets, Carrots, Horse-radish, Parsnips, Turnips, Potatoes, etc., in fall, they are put in temporary oblong heaps, on the surface of the ground on which they have been growing, and covered up with 5 or 6 inches of soil, which will keep off such slight frosts as are likely to occur until time can be spared to put them in permanent winter quarters; this is done in this section usually during the first part of December, in the following manner: A piece of ground as dry as possible is chosen; if not naturally dry, *provision must be made to carry off the water, lower than the bottom of the pit.* The pit is dug out from three to four feet deep, about six feet wide, and of the length required; the roots are then packed in in sections of about two feet wide *across* the pit, and only to the hight of the ground level. Between the sections, a space of half a foot is left, which is filled up with the soil level to the top; this leaves the pit filled up two feet wide in roots, and half a foot of soil, and so on until the whole is finished. The advantage of this plan is, that it is merely a series of small pits, holding from three to five barrels of roots, which can be taken out for market without exposing the next section, as it is closed off by the six inches of soil between. Also, that we find that roots of all kinds keep safer when in small bulk, than when large numbers are thrown into one pit together. In covering, the top is rounded so as to throw off the water, with a layer of from 18 inches to 2 feet of soil. This way of preserving roots, with perhaps the exception of Potatoes, is much preferable to keeping them in a cellar or root house, as they not only keep fresher, retaining more of their natural flavor and color, but far fewer of them are lost by decay than when exposed to the air and varying temperature of a cellar. Unmatured heads of Cauliflower, or Broccoli, however, are best matured in a light cellar or cold frame, by being planted in close together; in this way, good heads may be had to January. Cabbages are preserved very simply; they are left out as late as they can be pulled up by the roots, in this section about the end of November, they are then pulled up and turned upside down—the roots up, the heads packed close together, in beds six feet wide, with six feet alleys between, care being taken to have the ground levelled where the cabbages are placed, so that they pack nicely. They are left in this way for two or three weeks, or as long as the ground can be dug between the alleys, the soil from which is thrown in on the beds of Cabbage, so that, when finished, they have a covering of four or six inches of soil. This is not enough to cover the root however, which is left partly exposed, but this is in no way injurious. Some prefer to cover them up at once by plowing a furrow, shoveling it out wide enough to receive the heads of the Cabbages, then turning the soil in on the heads, and so continuing until beds of six or eight feet are thus formed. This plan is rather more expeditious than the former, but it has the disadvantage of compelling them to be covered up at once by soil, while the other plan delays it two or three weeks later, and it is of the utmost importance in preserving vegetables that the operation (particularly the final covering) be delayed as late in the season as frost will permit. Generally more are lost by beginning too soon than delaying too late. Onions, we find, are best preserved in a barn or stable loft, in layers of from 8 to 10 inches deep, covered up with about a foot of hay or straw on the approach of severe frosts. The great points to be attained are a low temperature and a dry atmosphere; they will bear 20 degrees of frost without injury, provided they are not moved while frozen, but they will not stand a reduction of temperature much lower than this without injury.

CHAPTER XVI.

INSECTS.

We have but little trouble with insects in our highly cultivated grounds; what with continued moving of the soil by plowing and harrowing every foot, from three to four times each season, incessant hoeing, and the digging up of the crops, we give these pests but little chance for a foot-hold. We are, however, occasionally troubled with *Aphides*, the "Green-fly," in our forcing houses of Lettuce. A complete remedy for this trouble, in its early stages, is smoke from burning tobacco stems; or tobacco stems steeped in water to give it about the color of strong tea, and applied with a syringe, will thoroughly destroy them. "Jumping Jack," or the Turnip-fly, occasions some trouble with late sowings of Cabbages, Turnips, and Radishes, but we find an excellent preventive in dusting lime over the beds, immediately the seeds begin to germinate. It is of the utmost importance to use *preventives* in the case of insects, for if once they get a lodge ment, it is almost useless to attempt their destruction. The striped Cucumber-bug, which, with us, attacks late sowings only, we have found to yield readily to a few applications of bone-dust, which serves the double purpose of disturbing the insect and encouraging the growth of the crop. But our most formidable enemy of the insect tribe is that which attacks the roots of the Cabbage family, causing the destructive disease known as the "club-root." There is a general misconception of the cause of this disease; happily our peculiar location here, gives me the means, I believe, of thoroughly disproving some of these absurd dogmas, that club-root is caused by "hog manure," "heavy soil," "light soil," etc. I do not doubt that it has appeared thousands of times with just such conditions; yet, within three miles from the City Hall of New York, I can show to-day, on the classic shores of Communipaw, scores of acres that have been just so manured, both light soils and heavy soils, that have grown Cabbages for twenty consecutive years, and yet, the first appearance of club-root is yet to be seen. On the other hand, I can show on soils, not more than a mile distant from those on the Communipaw shore, where the ground is cultivated in the very best possible manner, and where every variety of manure has been tried, and yet it is impossible to get a crop of Cauliflower or Cabbage clear from club-root for two years in succession. Now, the reason of the immunity from the pest on the one variety of the soil, and not on the other, does not, to us, admit of the slightest particle of doubt. On the shore side, and for nearly a mile inland, there are regular deposits of oyster shell, mixed with the soil, almost as we find pebbles in a gravelly soil; now, our theory is, that the insect which occasions the club-root, cannot exist in contact with the lime, which of course is present in large amount in a soil, containing such abundance of oyster shell. Reasoning from this, we have endeavored to bring up soils deficient in shell, by heavy dressings of lime; this answered, however, only temporarily, and we found it too expensive to continue it. The increasing demands for manures in the vicinity of New York, has rendered them of late years scarce and high in price, so that we were necessitated to begin the use of guano and other concentrated manures, and as this was rather new with us in our market gardens, we have had the pleasure of some very interesting experiments. Last season, in my grounds at Jersey City, where we have never been able to get two crops of Cabbages successively, without having them injured by club-root, my foreman suggested to me to experiment with a bed, of about half an acre, to be planted with early Wakefield Cabbage. One-half of this he proposed to manure at the rate of 75 tons per acre with stable manure, the other half with flour of bone, at the rate of 2000 pounds per acre; this was accordingly done in the usual way, by sowing the bone-dust on the ground after plowing, and then thoroughly harrowing in. During the month of May we could see no perceptible difference in the bed; but just as soon as our first hot days in June came, down wilted the portion that had been dressed with stable manure, showing a well-defined line the whole length of the bed, and, on pulling the plants up, we found that our enemy was at work, while in that portion that had been dressed by the bone-dust, not a wilted plant could be seen, but, on the contrary, the crop had most unusual vigor. This experiment has been to me one of the most satisfactory I ever tried; it still further proves, that this destructive insect cannot exist to an injurious extent in a soil impregnated with lime, and also proves, that we have a most effective remedy in this valuable and portable manure. The experiment was, however, to me rather a costly one; our past experience told us that there was no reason to expect that the portion, on which the stable manure was used, would not be attacked by club-root, as it had borne a crop of Cabbage the previous year, and nearly twenty years' working of that soil had shown that this crop could never be grown successively two years; but experiments, to be satisfactory, must be done on a scale of some magnitude, and although I lost some $200 by the difference in the crop, I believe it to have been a profitable investment.

I have incidentally stated that the Cabbage crop, treated in the usual manner, can only be grown every alternate year, the reason of which we infer to be, that the insect is harmless to the plant when in the perfect state the first season, but that it is attracted by the plant, deposits its eggs in the soil, and that in the larva condition in which it appears the second year, it attacks the root. Whether this crude theory is correct or not, I will not presume to say, but if it is not, how can we account for the fact of our being able to grow this plant, free from its ravages every alternate year, while, if we attempt to do so successively without the use of lime, it is certain to be attacked?

All authorities on gardening, that I have had access to, seem to be unaware of the fact that club-root is never seen in soils impregnated with shells. This variety of soil is not common. I have never seen it anywhere except here, and as I have before said, this peculiarity of location most fortunately gives a certain clue to the facts, and directly points out the remedy, which, I think, we have found to be in the copious use of bone-dust as manure.

CHAPTER XVII.

VEGETABLES, THEIR VARIETIES AND CULTIVATION.

In describing the modes of cultivating the different varieties of vegetables, I shall notice at length only those of the most importance, and the most profitable for market purposes, while for those of less value as market crops, the directions for culture will be such as are adapted to private gardens only.

A limited number of kinds will be described, and such only as our experience has shown to possess the greatest earliness and productiveness. Nothing is more perplexing to the beginner, than to be bewildered by descriptions of, perhaps twenty, so-called varieties of a vegetable, that perhaps, in reality, does not embrace four distinct kinds. For example, in early Cabbages, there are some hundred or more varieties described; yet we find, after having experimented with some scores of kinds in our time, there is one variety more profitable to grow than any other, viz. the Jersey Wakefield, which is grown in this locality to the exclusion of all others. However, some kinds are found to do better in some localities than in others, hence, as in fruits, no particular variety should be claimed to be *universally* the best.

ASPARAGUS.—(*Asparagus officinalis.*)

Asparagus being a hardy perennial plant, that may be grown on the same ground for twenty years without renewal, special care is required in forming the beds in which it is to grow. This is done sometimes by trenching to the depth of two or three feet, mixing each layer of soil, as turned over, with two or three inches of well rotted manure; but for market purposes, on a large scale, trenching is seldom resorted to; deep and thorough pulverizing by the plow and subsoiler serving instead. The soil best suited for Asparagus is a deep and rather sandy loam, such as is often to be found on the borders of meadows or on the margins of lakes—land formed by the washings of the higher grounds, and known as alluvial

VARIETIES.—There is considerable difference of opinion concerning varieties. Some contending that there are five or six, and others that there is only one variety, which is sometimes modified by differences of soil or climate. In this latter opinion we entirely coincide, believing that the *Asparagus officinalis* of our gardens is confined to only one variety, and that the so-called "Giant" can be made gigantic or otherwise, just as we will it, and that the "purple top" variety will become a "green top" whenever the composition of the soil is not of the kind to develop the purple, and *vice versa.* All practical gardeners know how differently soil and climate change the appearance of the same variety. Seeds of Cabbage, taken from the same bag and sown at the same time, but planted out in soils of light sandy loam, heavy clayey loam, and peat or leaf-mold, will show such marked differences when at maturity, as easily to be pronounced distinct sorts. This, no doubt, is the reason why the multitude of varieties, of all vegetables, when planted side by side to test them, are so wonderfully reduced in number.

Propagation.—Asparagus is propagated by seed which is sown in spring, as soon as the soil will admit of working, which should be prepared, by being thoroughly pulverized, and enriched with well-rotted manure. The seed is sown in rows 1 foot apart, and if kept carefully hoed, and clear from weeds, the plants will be in fine condition to plant out the succeeding spring. Strict attention to this will save a year in time; for if the seed bed has been neglected, it will take two years to get the plants as large as they would be in one year, if they had been properly cared for. In consequence of this very common neglect of proper cultivation of the seed bed, it is an almost universal impression that the plants must be two or three years old before planting. This is undoubtedly an error, for almost all large growers for market purposes, in the neighborhood of New York, invariably plant one-year old plants, and count on marketing a crop the third spring from the time of sowing. One pound of seed will produce about 3000 plants; and to plant an acre of Asparagus requires from 15,000 to 20,000 plants.

Planting.—The bed being prepared as previously described, planting may be done any time for six or eight weeks from the opening of spring; the plant, from its peculiar succulent roots, is less susceptable of injury from late planting than most other vegetables, although at the same time delay should not occur, unless unavoidable, as the sooner it is planted after the ground is in working order, the better will be the result. When there is plenty of ground, and the crop is to be extensively grown, perhaps the best mode of planting is in rows 3 feet apart, the plants 9 inches apart in the rows. For private use, or for marketing on a small scale, beds should be formed 5 feet wide, with three rows planted in each; one in the middle, and one on each side, a foot from the edge; the distance of the plants in the rows, 9 inches; the alleys between the beds should be 2 feet wide. In planting, a line is set and a cut made, a little slanting, to the depth of 6 or 8 inches, according to the size of the plants. The plants are then laid against the side of the trench, at the distance already named — 9 inches — care being taken to properly spread the roots. The crown or top of the plant should be covered about 2 inches. In a week or so after planting, the beds should be touched over lightly with a sharp steel rake, which will destroy the germinating weeds. The raking had better be continued at intervals of a week or so, until the plants start to grow, when the hoe may be applied between the rows and alleys; the weeds that come up close to the plant, must of necessity be pulled out by the hand.

The Application of Salt to Asparagus as a top-dressing, is of great benefit in inland districts, out of the range of a saline atmosphere; but is of little or no benefit in the vicinity of salt water. When used, it should be applied in spring only, at the rate of from 2 to 3 lbs. per square yard, strown on the surface; the rains will dissolve it and wash it down to the roots. Besides its beneficial effects upon the plant, it is destructive to the wire worm, and other insects that are often troublesome to the Asparagus.

We have found Asparagus beds very profitably benefited by the application of superphosphate of lime, as a spring top-dressing, applied at the rate of 500 lbs. per acre, sown on the beds and hoed in. Experiments with this, on alternate rows, showed a difference of nearly 1 foot in hight of the stalk, in favor of the rows to which the superphosphate had been applied, over those which had none; and a difference of nearly double the product when the crop was cut in the succeeding spring.

The fall treatment of the Asparagus beds varies with the locality; in cold regions, where, if left unprotected, the frost would penetrate below the roots, a covering of 3 or 4 inches of rough manure or leaves is necessary. Although an entirely hardy plant, it will start earlier, and with greater vigor in spring, if the root has not been subjected to severe freezing. In milder sections, no such precaution is necessary; all that need be done is to clear off the stems as soon as they are withered in the fall, and clean the beds preparatory to giving a dressing of 2 or 3 inches of manure, which had better not be applied until spring. We believe the common practice of top-dressing Asparagus beds in fall to be a very wasteful one, in districts where it is not necessary to provide against severe freezing, for, as the plant is then dormant, the juices of the manure are either evaporated, or else washed down by rains below the roots of the plant. I remember, many years ago, having three small Asparagus beds under my charge, on one of which I applied in December 25 lbs. of Peruvian guano, dissolved in fifty gallons of water; in April the same application was made to another bed, and the other was left without anything. There was no perceptible difference between that to which the liquid had been applied in December and that to which none had been given, but on that which received it in April, nearly double the weight of crop was produced. Since then, all our practice, corroborated by direct experiment, has convinced me beyond all doubt, that manures, either liquid or solid, organic or inorganic, are unprofitably employed when applied to plants in the dormant state.

In gathering the crop, caution must be used not to injure the plants by continued cutting; for it must be borne in mind, that to reproduce annually its crop of shoots in spring, something must be left to grow to encourage the formation of fresh roots. In our market gardens, the practice is to cut off all the shoots as they are ready, until the middle of May or 1st of June, when the shoots begin to show signs of weakness; then all is left to grow and no more cut. In its preparation for market, the shoots are tied up in round bunches, containing from twenty to thirty shoots in each. The tying material is usually bass-matting, as that is soft and has the necessary strength. It requires a little practice to do the "bunch ing" rapidly, and it should be the object of the beginne to strive to attain this, as it is light work, and continued slowness in the operation will make a serious gap in the profits.

This crop is subject to so many conditions, that an average value can hardly be given; some of our growers here claim that it pays an annual clear profit of $1000 per acre, while others say that it does not pay them over $200 per acre. During a period of ten years, counting from the time the bed was planted, it is safe to say that, in this vicinity, the average profits per acre will be $400. It is a crop that never fails to sell, is one that is always productive if it has been properly treated, and as it has a great value for its weight—a ton often being worth from $200 to $400—it is, in all respects, a valuable crop for the market gardener.

In some localities, especially on Long Island, the Asparagus-beetle has injured the crop to such an extent as to cause whole plantations to be plowed under. When the beetle first appears, it may be controlled; but if allowed to become established, the task is hopeless. The engraving, (fig. 26), shows the insect in its different stages. The lower figure is a part of a branch with the small black eggs attached by their ends; these are given of the natural size, and magnified. The larva, or caterpillar, as well as the perfect beetle, are shown at the top of the engraving; the natural size of these is indicated by the lines drawn at the side. Whenever the eggs or the larvæ appear, cut and burn the plants, as long as any traces of the insect are to be seen; this must be done if it destroys every vestige of vegetation.

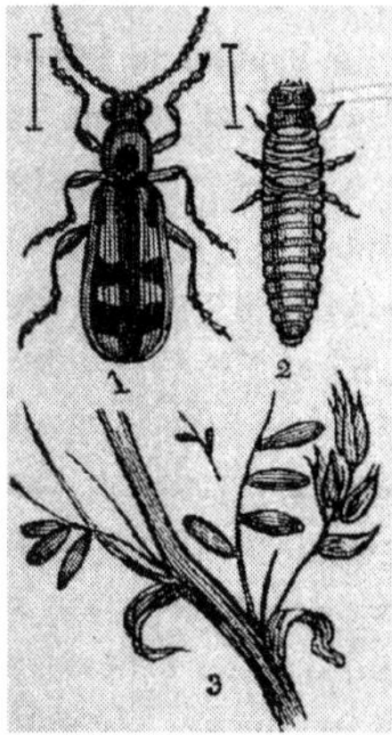

Fig. 26.—ASPARAGUS BEETLE.

ARTICHOKE.—(*Cynara Scolymus.*)

Although a vegetable as yet rarely seen in our markets, it is extensively used in Europe, particularly in France. The portion of the plant most used, is the undeveloped flower-head, or rather those portions of the flower-head called the scales of the involucre. They are sometimes boiled, and used as a salad, with vinegar, oil, and salt; but more generally in the raw state.

Fig. 27.—GREEN GLOBE ARTICHOKE.

Another use of the Artichoke is to blanch it, by tying the young side shoots moderately close together, as we tie Endive, filling in between with soil to exclude the air until the shoots are blanched; this is what is known as "Artichoke Salad," or "Artichoke Chard," it is used in this state in various forms of cookery, besides being used as a salad.

It is a vegetable of easy culture, originally propagated from seeds, until a stock is secured, after which it is readily increased by suckers from the root. These are planted out in April or May, in rows from 3 to 4 feet apart, and 2 feet between the plants, care being taken that the plants are well firmed in planting, and if the weather is dry, they must be freely watered until they start to grow. The plantation, the first season, will only give a partial crop; but, as it is a perennial plant, after being once planted, the same bed will remain in bearing for years. The plant may be said to be entirely hardy south of Mason and Dixon's Line, but north of that, it requires to be protected by covering between the plants, with 6 or 8 inches of leaves or coarse manure.

THE VARIETIES are the Green Globe, and Common Green, differing but little, except in the form of the flower-bud, the former being globular, the latter conical. It is claimed by some that the Common Green is more hardy and productive, but we have grown them side by side for years, and never have observed any difference, except the very trifling one in the shape of the flower-bud.

ARTICHOKE.—JERUSALEM—(*Helianthus tuberosus.*)

This is an entirely different plant from the true Artichoke, though it resembles it somewhat in flavor—hence its name. As it is very often confounded with the true Artichoke, we give an engraving of both. This one is a species of *Helianthus*, or Sun-flower, and the plant has the general appearance of a small Sunflower. The edible part of the plant is its tubers. These are like the potato in appearance, but when cooked, to the taste of most people, are very inferior in flavor. Their nutritive value is said, however, to be fully equal to that of the potato. Used in the raw state, it is pickled like the cucumber, or sliced, and eaten with vinegar as a salad, but as a culinary vegetable is but little grown, except for variety or novelty.

Its culture is, in all respects, similar to the potato, but it is more productive, always free from disease, will grow almost in any soil or situation, and will stand the winter on light soils wherever a Parsnip crop will stand; for

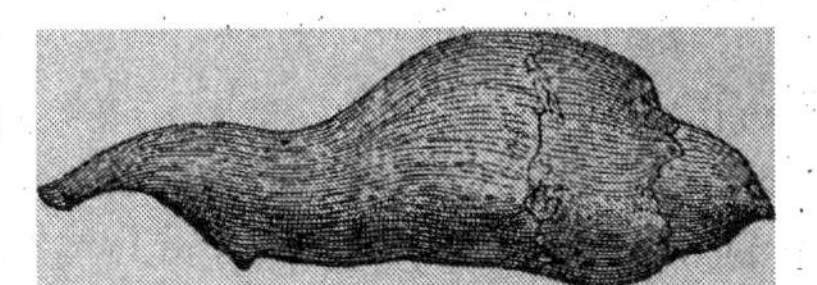

Fig. 28.—JERUSALEM ARTICHOKE.

these reasons it has been suggested that it might prove a valuable food for cattle, or pigs, who eat it as freely as potatoes, when boiled.

There are several varieties known as *Red*, *Purple*, *Yellow*, and *White Skinned.*

BASIL.—(*Ocimum Basilicum.*)

An herb of a highly aromatic odor, and a strong flavor of cloves. It is used for flavoring soups, stews, and sauces, and is by some used in salads. Its culture is the same as that of other sweet herbs. The seed should be sown in the *open ground*, and not in frames, which is the English practice, and necessary there from their colder climate. Sow in rows 1 foot apart; when 3 or 4 inches high, it may be transplanted in rows 1 foot apart, and 6 inches between plants. If a small quantity only is required, it may be thinned out in the seed rows, and left to grow where sown.

There are two species cultivated, namely, the Common Sweet Basil, (*Ocimum Basilicum*), and the Bush Basil, (*O. minimum*).

BALM.—(*Melissa officinalis.*)

Another well-known aromatic herb, which has a very agreeable lemon-like odor. It is used as a tea for its soothing effect in irritations of the throat and lungs, and a century ago was used as a specific for coughs and colds. Its young shoots are sometimes used as an ingredient in salads. It is rapidly propagated by divisions of the root, which, planted in spring, at 1 foot apart each way, will form a solid mass by fall.

Besides the common kind, we have now in cultivation a beautiful variegated variety, possessing all the properties of the other.

BEAN.—(*Phaseolus nanus*)—Bush, Kidney, or Snap.

A leading vegetable of our market gardens, and extensively cultivated in every section of the country, North and South. Although it can be grown on soils that are not enriched by manure, yet, like almost every other vegetable, it is more profitable when grown on highly cultivated land. It is, what we term, a "tender" plant; that is, one that will be killed by the action of slight frosts, hence it is not planted until late enough in the spring, to secure it from the risk. As in a country presenting such differences of temperature as ours, no stated date can be given at which to sow, perhaps no safer rule can be adopted for sowing all "tender" vegetables for all parts of the country, than the time at which our great staple, Indian Corn, is planted. This rule will be equally intelligible to the inhabitants of Maine and to those of South Carolina, for all plant Corn and know, that our great enemy to early vegetation, "Jack Frost," will, without scruple, smite this "tender" vegetable if it be forced to grow before his icy reign is past. In this section, we plant Beans for *first crop* when we plant Corn, from 10th to 20th May. But as the crop of Beans comes rapidly to maturity, under favorable circumstances, in five or six weeks, it may be sown any time from these dates until July, August, or September, according to the temperature of the district.

The culture in market gardens, is simply to draw drills about 3 inches deep, and from 18 inches to 2 feet apart, according to the richness of the soil; the poorer the soil, the closer they can be planted. The seed is dropped in the drills 2 or 3 inches apart, and the soil covered in on them with the feet; this we find to be a quicker and better method of covering in seeds of this size, than by the hoe or rake. After the plants have grown an inch or two, a cultivator is run between the rows, which generally is all that is necessary to be done, until they are large enough to have a little earth thrown to each side of the row by the plow, which completes their cultivation. Beans, like Tomatoes and Peas, are easy of cultivation, and not at all particular to soil, and are, in consequence, rarely a profitable crop *in the locality in which they are grown;* hence the only way in which they are made profitable is, by growing South and shipping North, they being easily transported. Large quantities are grown in early soils in southern sections of the country, and shipped to our large northern cities, and meet a rapid sale, at prices that must pay a large profit, if their manner of growing the crop is as simple as ours. It will be understood that thi crop is almost exclusively sold in pod, as snap shoots, (in the green unripened state), by the hucksters, and rarely as a shell bean.

The varieties are now very numerous, but the following, placed in what we consider the order of their value to the market gardener, will embrace variety enough for all practical purposes.

Early Valentine.—Early, productive, tender, succulent, and of excellent flavor; continuing longer in the green state than most of the varieties. Seeds, when ripe, salmon, speckled with purplish-rose. This variety is often marketable in six weeks from the time of sowing in May.

China.—Rather earlier than the preceding, but hardly so productive; the pods become yellow quicker, which makes it not so valuable as a market variety. It is, however, grown by some in preference to the Valentine, it being considered a few days earlier. Seeds, when ripe, white purple, speckled.

Mohawk.—This variety is the most suitable for northern latitudes, as it is less susceptible of injury from cold than most of the others; it is very productive, with pods 5 or 6 inches long, but is not recommended as an early variety. Seeds drab, variegated with purple spots.

Refugee, or Thousand to One. — Very productive, though not early; young pods extremely tender and of fine flavor. This variety is very extensively grown for pickling, and has long been a standard sort. Seeds, dull yellow, speckled with purple.

Newington Wonder.—A wonderfully productive sort, and one of the most popular of all varieties for private use, the pods being particularly crisp and tender; the most valued forcing variety. Seeds light brown, lined with yellow.

White Marrowfat.—This variety is the one so extensively grown for sale in the dry state; it is also valued as a string bean, but is used to greater extent shelled, either green or dried. Seeds large, ivory white.

Turtle Soup.—This receives its name from some fancied resemblance that soup made from the ripe beans, has to that made from the turtle. It is a late variety, requiring the whole season in the Northern States to ripen its seeds. Seeds small, glossy black; generally used when ripe.

BEAN.—(*Phaseolus vulgaris, etc.*)—RUNNING, OR POLE.

These require rather more care in culture than the Bush Beans. The soil best suited is sandy loam, which should be liberally enriched with short manure in the hills, which are formed, according to the variety, from 3 to 4 feet apart, and provided with a stake from 8 to 9 feet in hight, set in the centre of each. This class of beans is particularly tender, and it is perfectly useless to plant the seeds before the weather has become settled and warm in spring, as they are almost certain to rot, and even should they not, the plant makes no growth, unless in uninterrupted warm weather. In this district, they should never be planted out until a week or more after the planting of Bush Beans; if planted at the same time, 10th of May, they are almost certain to fail.

From five to six seeds are planted in each hill, about 2 inches deep. Being a vegetable requiring considerable expense in growing, staking, picking, etc., it brings a correspondingly high price per acre in market; but maturing during the heat of summer, the advantage of earliness in this crop is not so manifest as in many others. The profits per acre average about $250, when sold for consumption in the locality in which it is grown. Shipped from earlier sections it, no doubt, would double the above amount. There are many varieties, but only a few of leading value.

Lima.—(*Phaseolus lunatus.*)—This is almost universally grown both for market, and for private use. It is esteemed the best of all the pole beans.

Small Lima, or Sieva.—A variety of the preceding, differing in flavor from the common Lima, and by some much preferred. Habit of the plant similar. Seeds white; very productive.

Asparagus Bean.—This variety is most suitable for a warm climate, as it takes a long season to grow; pods, when full grown, are from 12 to 15 inches long; they are used as string beans, or for pickling in the green state; it is not used as a shelled bean, being much inferior to the Lima.

Dutch Case Knife.—A very productive variety, with long and broad pods; it is of excellent flavor, and next to the Lima, is the best market sort.

5*

London Horticultural.—A very popular variety for private use, as it is equally serviceable in the green state, or, when mature enough, to shell. Seeds oval, marbled with purplish-brown.

Scarlet Runner.—(*Phaseolus coccineus.*)—A great favorite in European gardens, both as an ornamental plant, and a useful vegetable. It grows to the hight of 9 or 10 feet, producing dazzling scarlet flowers, from July to October. Used as string beans, and shelled. Seeds lilac, mottled with black.

White Runner.—Similar in all respects to the above, except in color of flowers and seeds.

Red and White Cranberry.—These are intermediate in season of maturing. Very popular sorts, used either as string beans or shelled.

BEET.—(*Beta vulgaris.*)

This is one of the leading and most valuable crops of our market gardens, and next to Cabbages, is perhaps the most extensively grown as an early crop. The soil best suited, is that which is rather light than otherwise, always provided that it is thoroughly enriched by manure. We make little difference in the manner of working or manur ing the ground for any of our leading early crops; the ground must, in all cases, be thoroughly pulverized by plowing, subsoiling, and harrowing, and when stable manure can be procured, plowed in at the rate of 75 or 100 tons per acre. If stable manure cannot be had, the next best substitutes must be used in the quantities specified·

see Chapter on Manures. As early in spring as the ground becomes fit to work, the Beets are usually sown in rows 1 foot apart, made by the "marker," about 3 inches deep. We prefer to sow rather thickly, not less than 8 lbs. per acre, for the reason that late frosts often kill off a portion of the young plants, but when sown thickly, enough is generally left to make a crop, which amply repays the difference of a few pounds of seed. After sowing, the drills are covered in by the feet, by walking along the rows, after the bed is completed; if the weather is dry, the whole is rolled, which better firms the soil around the seed and also leaves the ground level, making it easier to be hoed. Beets are occasionally planted 2 feet apart, and the intervening row sown with Radishes; the Radishes mature early, and are used or sold off soon enough to admit of more room for the Beet crop. It makes with us but little difference in the profits of the crop which way it is done, the results being nearly the same in each case; but in places where limited quantities of vegetables only can be disposed of, perhaps the latter plan is the best. The young Beets are thinned out to 6 inches apart when the rows are 1 foot apart, but when at 2 feet to only 4 inches, as they have more space between the rows for air. The thinnings of the Beets are used like Spinach, and when carefully handled, the thinnings will always sell for more than the price of the labor of thinning the crop.

In this neighborhood, Beets sown first week in April, are begun to be marketed the first week in June, and entirely cleared off by July 1st, when the ground is prepared for the second crop. It will be understood that they are at this early date sold in an immature state, before the root has reached complete development, but the great point is earliness; the public being well satisfied to pay more for it half-grown, if early, than when full grown, if late.

This crop I have always considered a very profitable one, even at the seemingly low price of $1 per 100 roots, the average wholesale price in New York markets. But 80,000 roots are grown per acre when sown at 1 foot apart, and although the labor of pulling and bunching up is greater than in some crops, yet, at $1 per 100, it will give an easy profit of $400 per acre.

Beets are an excellent article to ship, and the price paid in New York, for the first lots from Savannah and Norfolk, etc., is often as high as $3 per 100 roots.

The foregoing all relates to the crop in the green state for an early market, but they are also extensively grown for use in fall, winter, and spring. For this they are usually sown later, often in some sections as a *second* crop, as late as July 1st, although in the Northern States the roots hardly develop enough when sown after June. The manner of saving them in winter, will be found under the head of Preserving Vegetables in Winter.

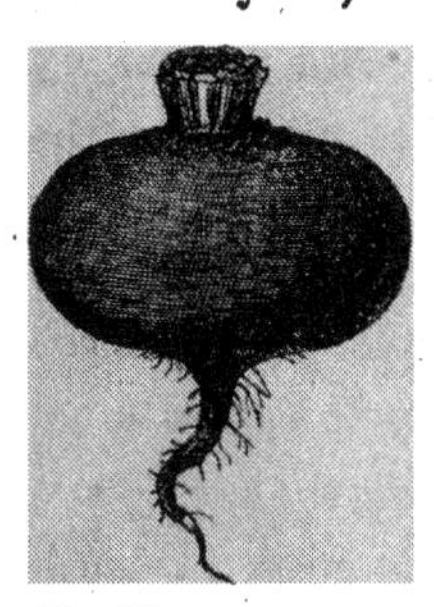

Fig. 29.—SHORT-TOP ROUND BEET.

The really useful varieties of Beets are very limited in number, and are embraced in the following, arranged as usual, according to their merit as market sorts.

Short-top Round.—This variety originated with us about ten years ago; it differs from the common Blood

Turnip Beet, in being rather flatter, freer from roots, and what is of main importance, shorter in top; it is not quite so early as the Bassano, but being of richer color, it at once supplants it in market, soon as it comes in, which is usually in three or four days after that variety.

Bassano.—The earliest of all known varieties; outside color light red; flesh white, veined with pink. Its earliness is its only merit, as it is coarser than the deep colored varieties. The proportionate quantity sown for market purposes, should not be more than one-sixth of the preceding.

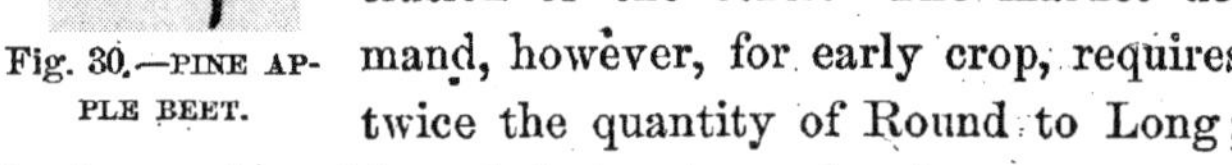

Fig. 30.—PINE APPLE BEET.

Pine Apple.—An excellent variety of rich deep crimson color, pine apple shaped, and nearly equal in earliness to the Short Top Round.

Long Smooth Blood.—A great improvement on the Common Blood Beet, being less strong and freer from rootlets, besides being a week earlier. It is now grown here to the entire exclusion of the other. The market demand, however, for early crop, requires twice the quantity of Round to Long; for late sales of barreled roots, exactly the reverse.

Swiss Chard.—Is a distinct species from the Beet grown for its roots, known to botanists as *Beta cicla.* It is cultivated solely for its leaves. The mid-rib of leaf is stewed as Asparagus, the other portions of the leaf being used as Spinach. The outer leaves are pulled off as in gathering Rhubarb. It is largely grown in France and Switzerland.

In this country, it is now cultivated to some extent in private gardens only. Its handsome foliage is as attractive as many of our prized flower-garden "leaf plants," and no doubt it would be much valued if we could only regard it without the idea that it is only a Beet.

BORECOLE OR KALE.—(*Brassica oleracea. Var.*)

A variety of this receiving the rather indefinite term of "Sprouts," is extensively grown for the Northern markets, many acres of it being cultivated in the vicinity of New York. It is sown in the month of September, in rows 1 foot apart, treated in every way as Spinach, and is ready for use in early spring. It is difficult to keep in some soils in winter; those of rather a light nature being the best. When successfully wintered over, it is a very profitable crop, not unfrequently selling for $500 per acre. The variety thus grown, is known in the seed stores as Dwarf German Greens. Another class of it is cultivated as we grow late Cabbage; it is sown in the open ground in May, and planted out at distances, according to the variety, from 2 to 3 feet apart. Of all the Cabbage tribe this is the most tender and delicate, and it is surprising that it has never yet been wanted in quantity enough to make it a marketable vegetable, not one head being sold to one thousand of the coarse winter Cabbage. The varieties are very numerous; those below described are all standard sorts.

Dwarf German Greens, or "Sprouts."—Color blueish-green, slightly colored, resembling somewhat the foliage of Ruta Baga Turnips; it is of delicate flavor, and every way desirable. The popular market sort.

Green Curled Scotch.—A rather dwarf variety, rarely exceeding 18 inches in hight, but spreading, when under good cultivation, to 3 feet in diameter. The leaves are

Fig. 31.—GERMAN GREENS.

beautifully curled, and of a bright green. This variety is very hardy, and will remain over winter in any place where the temperature does not fall below zero; it is most tender after being touched by sharp frost.

Purple Borecole.—Similar to the above in all respects except color, which is of a dull purple. It is the variety most esteemed by the Germans; it is very hardy, and is often seen in the markets of New York as late as January.

Brown Borecole.—Leaves brown, as the name indicates; merely a sub-variety of the purple, being in all respects the same, except in color.

Cottagers' Kale.—A comparatively new variety, a great favorite in England. It is dwarf, not exceeding 12 inches; leaves rich green, double curled and "feathered" almost to the ground. Very hardy, and a most profitable sort, more weight being grown in the same space than with any other variety.

BROCCOLI.—(*Brassica oleracea. Var.*)

This vegetable is so closely allied to Cauliflower, that it seems absurd to have ever divided them under different heads. Still we persist in growing them under the names of Broccoli and Cauliflower, the Broccoli being planted for fall use, Cauliflower, on the other hand, being mostly planted for summer use, although it is well known that their seasons might be reversed without any marked difference in the results. Like all of the Cabbage tribe, Broccoli, to grow it in perfection, requires the soil to be in the highest possible degree of fertility. The seed should be sown, in this district, in the early part of May, which will give plants large enough to be transplanted in July. Farther south the sowing should be delayed until June or July, and the transplanting delayed accordingly until August, September, or October. There is no doubt that in parts of the country where the thermometer does not fall below 20° or 25°, that Broccoli may be had in perfection from November until March. A necessary condition of perfect development, is a moist and rather cool atmosphere; for this reason we only get the crop in fine condition, in this district, during the cool and moist months of October and November. Owing often, however, to heat or dryness in the months of August and September, the crop becomes an entire failure, and for this reason, for market purposes, it is rather hazardous. When a good crop is made, however, it is very profitable, rarely bringing less than $12 per 100, or about $1000 per acre. The plants of most of the varieties are planted 2½ by 1½ feet, or about 10,000 plants per acre.

In this district, for market purposes, we confine ourselves to the first two varieties named below; the other two, however, are occasionally grown for family use.

White Cape.—Heads of medium size, close, compact, and of a creamy white color; one of the most certain to head.

Purple Cape.—Nearly similar in all respects to the White Cape, except in color, which is greenish-purple. This variety is rather hardier than the preceding, but its color renders it of less value in market. White heads of the same quality bringing $1 to $2 more per 100. This is mere matter of fancy in the buyers, however, as when cooked there is but little difference in its appearance from the White, and none whatever in the flavor.

Early Walcheren.—This variety seems to produce its heads earlier than the preceding, but they are not usually so heavy or compact.

Knights' Protecting.—This variety is of dwarf habit, much esteemed by private gardeners for preserving in frames or cellars, for late winter use. When lifted and

planted in boxes of earth in a cellar or in cold pits or frames, even as late as November, fine heads may be had until January.

BRUSSELS SPROUTS.—(*Brassica oleracea. Var.*)

This vegetable has never come into general use in this country, probably owing to its being too tender to stand the winters of the Northern States. Still, by sowing in April or May, and planting out in July, it may be had in fine condition until December; and in the Southern States, may be had in use from November to March. Even in England, where it is very extensively grown, it is not much raised for market, being mainly cultivated for private use. Its cultivation is very simple, and it can be grown on almost any soil. Plant about 2 feet apart, and cultivate as for Cabbages.

Fig. 32.—BRUSSELS SPROUTS.

There is only one kind, which is distinguished from all other varieties of the Cabbage tribe by the sprouts or buds, about the size of walnuts, which grow thickly around the stem; these sprouts are the parts used, and are equal in tenderness and flavor to Cauliflower or Broccoli.

CAULIFLOWER.—(*Brassica oleracea. Var.*)

As we remarked in the chapter on Broccoli, Cauliflower is mostly grown as a spring or summer crop, and as all such are more profitable, and consequently of more interest to the grower than crops maturing in fall, we will give its culture more at length.

Any soil that will grow early Cabbages, will grow Cauliflower, as their requirements are almost similar; but as the product is more valuable, extra manuring and preparation of the soil will be well re-paid. In situations where irrigation could be practised, it would be of great benefit in dry weather. We have occasionally found, when our beds were convenient to water, that even watering by hand has been of advantage. But few or no other crops of our gardens will re-pay that labor. The seeds of Cauliflower, (that we wish to be ready for market in June), are sown in the fall previous, between the 10th and 20th of September. In the course of four or five weeks the plants are transplanted into frames, in the manner described in the chapter headed "Uses and Management of Cold Frames;" but as they are rather more tender than Cabbage or Lettuce plants, we find it necessary to have the glass protected by straw-mats at night during winter. In cases where it is not convenient or practicable to have the plants thus wintered over, they can be had nearly or quite as well by sowing the seed in the hot-bed, or vegetable forcing house, in January or February, and transplanting the plants at 2 or 3 inches apart in boxes or in the soil of another hot-bed, until such time as they are safe to be planted in the open ground, which, with us, is usually from 15th of March to 10th of April. If properly hardened off, they are rarely

injured by being planted out too early. It must be borne in mind that the plant is nearly hardy, and that 10 or 15 degrees of frost will not injure it, provided it has been treated as its hardy nature requires, by having been exposed to the air previous to setting out in the open garden. I know that the general practice with amateur horticulturists is very different from this, and that their usual time of planting Cauliflower is when they plant Tomatoes, and in consequence, failure is almost universal. The plant, set out in May, hardly gets root before hot weather sets in, and if the flower head is developed at all, it is merely an abortion of what it should be. With me, for the past four or five years, Cauliflowers have been one of my most profitable crops. I have during that time grown about one acre each year, which has certainly averaged $1500. On one occasion the crop proved almost an entire failure, owing to unusual drouth in May; while on another occasion, with an unusually favorable season, it sold at nearly $3000 per acre.

The average price for all planted is about $15 per 100, and as from 10,000 to 12,000 are grown to the acre, it will result in nearly the average before named—$1500 per acre. Unlike Cabbages, however, only a limited number is yet sold, and I have found that an acre of them has been quite as much as could be profitably grown in one garden. Cauliflowers require careful handling to be marketed in good shape; after being trimmed of all surplus leaves, they are packed in boxes holding about 100 each, and are generally sold to retailers in this shape, without being removed from the packages. This early crop is always sold by the first week of July, allowing plenty time to get in second crops of Celery, etc.; but when wanted for fall or winter use, its treatment is the same in all respects as that of Broccoli. Like all our market garden products, we grow only a very limited number of varieties, and these such as are suited to our climate here; some of the most popular English sorts being perfectly worthless with us.

Early Erfurt.—This is our favorite sort, being a dwarf compact growing kind, producing uniformly large heads; the leaves grow more upright than in any other variety, consequently it can be planted closer, 24 inches by 15 inches, while most of the other sorts require 28 inches by 18 inches. This variety is comparatively new, and the seeds are very scarce and high priced.

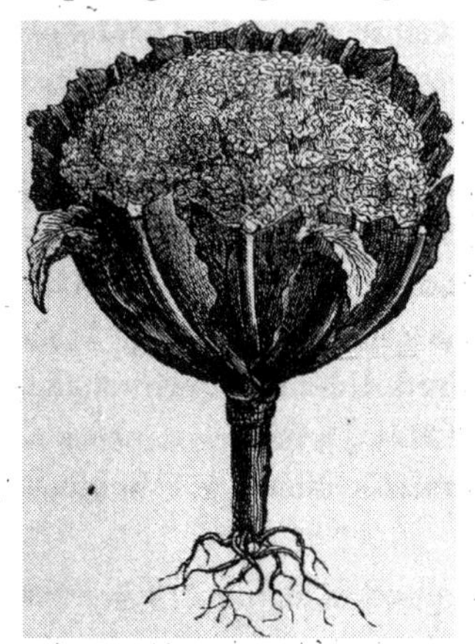

Fig. 33.—EARLY ERFURT CAULIFLOWER.

Early Paris.—This well-known variety stands next on the list; it is equally meritorious in all respects to the Erfurt, except that it requires more space to grow in.

Half Early.—A variety that is very useful for a succession crop. The great difficulty with Cauliflowers for market is, that the whole crop comes in and must be sold in the space of two weeks, unless we have varieties that come on in succession.

Wellington.—Recently introduced; forms a flower head of immense size; we have measured them 13 inches in diameter. It is, however, a large foliaged variety, and would require considerable space to grow in; for this reason it will not likely become a popular market variety.

CABBAGE.—EARLY.—(*Brassica oleracea.*)

The early varieties of Cabbage are cultivated more extensively than any other vegetable we grow. If they do not occupy a larger number of acres, they certainly sell for a much larger amount than any other crop. They are also generally considered to be the most profitable of all crops of our gardens on congenial soils. Experience in a great variety of soils in the cultivation of this crop, shows that what is known as heavy sandy loam, overlaying a porous subsoil, is the best adapted to it. Along the sea shore, for about one mile inland, we have often an admixture of oyster and other shells in the soil; wherever such is found, there, with proper cultivation, Cabbage can be raised in the highest degree of perfection. The large amount of lime in the soil, produced by the gradual decay of the shell, is not only congenial to the growth of the Cabbage tribe, but is certainly destructive to the larva of the insect which is known to produce club-root. In such soils, where in some instances Cabbages have been grown for fifty consecutive years, club-root is never seen. It is plain from this then, that lime is indispensable in the cultivation of this crop, and that if not naturally found in the soil, it must be applied. The most profitable application, I have found to be the flour of bone; a detail of some experiments with which will be found in the Chapter on "Insects."

The preparation of the ground for Cabbage differs in nothing from that for all the regular market garden crops, —careful plowing and subsoiling, and manuring with stable or barn-yard manure when procurable, at the rate of 75 tons per acre, alternating this with guano, etc., in the quantities named under the head of "Manures."

The early varieties of Cabbage are planted out in spring, as soon as the ground is dry enough to work; in the latitude of New York, from 15th March to 15th April. The distance apart is from 24 to 28 inches between the lines, and 16 inches between the plants in rows. At the same time that we plant out Cabbage or Cauliflower, between the lines are planted Lettuce, at 12 inches apart. To repay such expensive manuring and cultivation, every inch must be made to tell.

The Lettuce is ready for market by the middle of May, and is cut off before the Cabbage is large enough to injure it. The ground is now clear of the Lettuce, and the whole space is occupied by the Cabbages, which are all sold off before the middle of July; the greater part in fact by the end of June, which gives the necessary time for second crops. The raising of the plants is done in various ways, according to the differences of climate, and also of the market requirements in different sections. In latitudes where the thermometer never indicates 20° below the freezing point, Cabbage plants may be sown in the open border in October, and planted out at the distances named, on the first opening of spring; but in our Northern States, they must either be sown in hot-beds in February, (see instructions in Chapter on Hot-beds) or what is still better, wintered over in cold frames. For this purpose the seed is sown from the 10th to the 20th of September; strict attention to date is important; if too soon, the plants might run to seed, and if too late, they would be too small. In about four or five weeks from the time of sowing, they will be fit to transplant into the cold frames, from 500 to 600 being put under a sash 3 by 6 feet. In planting, it is very important with Cabbage or Cauliflower, that the plant is set down to the first leaf, so that the stem or stalk is all under ground, for we find that if exposed, it will be split by the action of the frost, and will be injured in consequence. Instructions regarding winter treatment, will be found under the head of "Cold Frames." I have before stated that, from the extent to which Early Cabbage is cultivated, it is one of the most important crops grown. It is also by far the most profitable, on a large scale, and no grower here, whose ground has not been fitted to produce it properly, has ever been very successful; it requires but little labor, and is always readily disposed of at profitable rates. At the distance planted, from 12,000 to 13,000 are grown per acre, the average price of which, at wholesale, is $50 per 1000, or about $600 per acre.

We allude to varieties here with some hesitation, as it is unquestionable that soil or climate has much to do in determining the merits of varieties in different localities As the best that I can do in the matter, I adopt the usual plan I have adhered to throughout, and place first on the list those we find to have the greatest general merit.

Jersey Wakefield.—This variety is said to have been first grown by Francis Brill, then of Jersey City, N. J., some twenty-five years ago, from a package of seed received from England under the name of Early Wakefield, and has been kept in the immediate locality almost ever since. We have experimented with scores of varieties in that time, and find nothing equal to it. It is quite a shy seeding sort, and on several occasions enough seed could not be procured to meet the demands of the growers, and then it has repeatedly sold as high as $20 per lb., or quite five times the rate of other sorts. True, we have it quoted in the English seed lists as low as other varieties, but repeated trials of almost every kind named in their catalogues, too well told us that the Early Wakefield, as we know it, was no longer procurable in England. The merit of this variety consists in its large size of head, small outside foliage, and its uniformity in producing a crop. The heads are pyramidal, having rather a blunted or rounded peak; color glaucous green.

Fig. 34.—JERSEY WAKEFIELD CABBAGE.

Early York.—This well-known variety is more universally cultivated than all others; in earliness it is quite equal to the Wakefield, but is inferior in size, and for market purposes, with us, would not sell for much more than half the price of the Wakefield. Heads small, roundish-oval; color pea-green.

Large York.—Similar to the above, but larger in all its

parts; it is grown to a considerable extent in the South and South-west.

Early Ox Heart.—This, next to the Wakefield, used to be our favorite sort for market purposes, and although equal in earliness and size, was found not be so uniform in heading; for family culture it is, however, a very valuable variety, as it is one of the best flavored and tender.

Early Winningstadt.—Should hardly be claimed as early, as it is quite three weeks later than any of the above, but it is an excellent sort where earliness is not an object, as it heads uniformly, and is of large size, often weighing 20 lbs. It is a very distinct variety; head pyra-

Fig. 35.—EARLY FLAT DUTCH CABBAGE.

midal; the outer leaves spiral and spreading, which requires it to be planted wider than the early sorts. For this reason, together with its lateness, it is not a favorite in gardens where two crops are grown in one season.

Early Flat Dutch.—A very dwarf variety with large round head, almost flat on the top; it is a very excellent variety for a succession crop, being two or three weeks behind the earliest sorts. Though not more than 8000 or 9000 can be planted on an acre, yet, as it comes in just when the glut is over, it rarely sells for less than $12 per 100. Its lateness, however, prevents the getting in of a second crop, and it is consequently not largely grown.

CABBAGE.—Late.

The manner of cultivating Late Cabbage is not quite so expensive as that for Early, and as a consequence, the receipts for the crop are correspondingly low. In fact, it is often sold at prices that would not more than repay the price of manure and labor expended on the early crop. But as it can be raised with much less manure and labor, and on land less valuable, it is extensively grown in the neighborhood of all our large cities, rather, however, by farmers than by gardeners. The seed is sown usually in the early part of May, and the plants set out in July, at distances of 3 feet between the rows, and 2 feet between the plants. The crop is almost exclusively worked by the cultivator or plow, one hoeing usually sufficing around the plants. In Long Island, N. Y., they are set out in July, on the ground from which early Potatoes or Peas have been grown. About 10 tons of stable manure per acre is usually put in the rows over which the plants are set. The price averages about $40 per 1000; 6000 or 7000 are grown per acre, giving an average of, perhaps, $300 to the acre. Late Cabbage is extensively shipped during the fall months, from New York to southern ports. The hot and dry summers there preventing the raising of plants from seed. Recently, however, some of the growers in Charleston, Savannah, Richmond, and other cities, have discovered that it is more profitable to have the plants grown North, and to plant them in August or September, and grow them themselves. Many hundred thousands of plants of Cabbage, Cauliflower, and Celery, are now annually sent South in August.

The varieties of Late Cabbage are not so numerous as the Early; the best for general purposes are the following:

Bergen Drumhead.—This is the variety grown for the general crop, it is of the largest size, sometimes almost round, though more generally flattened at the top. It is extremely hardy, and will withstand severe frosts without injury. In localities where there is not more than 15 or 20 degrees of frost, it can be left out where grown all winter, but in the Northern States requires the protection as recommended in the chapter on "Preservation of Vegetables in Winter."

Premium Flat Dutch.—A very handsome variety, differing from the Drumhead in perfecting its head rather earlier in the fall, and for that reason is not quite so well adapted for winter use; it is, however, much grown as an early fall sort. It is particularly tender, and superior in flavor to the Drumhead.

Mason.—Sometimes called Stone Mason, in compliment to its extreme hardness, I suppose. Is rather a small variety for a late Cabbage, but this enables it to perfect its head in a short season, and for this reason it is recommended for extreme northern latitudes.

Drumhead Savoy.—This variety is the largest of the Savoy Class, and is the sort most generally cultivated for market. The head is large, spherical, very solid and compact, of a yellowish-green; and like all others of the Savoy varieties, is of excellent flavor, far surpassing that of any late Cabbage. Still, such is the force of habit, that the public do not purchase one Savoy for every thousand

Fig. 36.—DRUMHEAD SAVOY CABBAGE.

of the coarse Drumhead class, although the difference in quality between the two is as great as between the fox grape of the woods and a cultivated Delaware.

Green Globe Savoy.—Smaller in all respects than the preceding, of darker green, the leaves intensely wrinkled. The compact and rather upright growth of the lower leaves allows it to be planted quite as close as early Cabbage, 24 inches by 18 inches. It is the favorite of all the varieties for family use.

Red Dutch.—Is used almost exclusively for pickling;

it is one of the hardiest of all Cabbages, and when preserved as directed for the others, will keep later in the season than any other. It is slow to mature, however, and requires a richer soil for its perfect development.

CARDOON.—(*Cynara cardunculus.*)

A vegetable that is but little grown, and then oftener as a novelty than for use. It belongs to the same family as the Artichoke, which it much resembles. The shoots, after blanching, are used in soups or in salads. It is cultivated by sowing the seeds in early spring, thinly, in rows 3 feet apart, and thinning out to 18 inches between the plants. The plant attains its growth in early fall, when it is blanched by tying the leaves together so as to form an erect growth, after which it is earthed up, and preserved exactly as we do Celery.

CARROT.—(*Daucus Carota.*)

This may be classed more as a crop of the farm than of the garden, as a far larger area is grown for the food of horses and cattle than for culinary purposes. Yet it is a salable vegetable in our markets, and by no means an unprofitable one to grow on lands not too valuable. It is not necessary that the land for this crop should be highly enriched. I have grown on sod land, (which had been turned over in fall), 300 barrels per acre, without a particle of manure, and three years after, the same land which had been brought up to our market garden standard of fertility, a very inferior crop; the land being too rich, induced a growth of tops rather than roots. In our market gardens, we sow in rows 14 inches apart, thinning out to 3 or 4 inches between the plants; but on farm lands where space is not so valuable, they should be planted 18 or 24 inches between the rows, and worked with the cultivator. For early crops, we sow at the beginning of our first operations in spring, in the same manner as we sow Beets, as soon as the ground is thoroughly dry; but for later crops, they may be sown any time, in this latitude, until the middle of June. This is one of the vegetables that require a close watching, to see that it does not get enveloped with weeds, as in its early stage it is of comparatively feeble growth, and unless it is kept clean from the start, it is apt to get irrevocably injured.

The usually prescribed quantity of seed per acre is 5 lbs., but I have always considered it safer to sow nearly double that quantity. In dry weather it germinates feebly, and not unfrequently, when seed comes up thinly, it is scorched off by the hot sun, and the saving of a few pounds of seed may entail the loss of half the crop. We prefer to sow all such crops by hand.

The Carrot, like all other root crops, delights in a sandy loam, deeply tilled. Considerable quantities of the early varieties are sold, in our markets, in bunches, in a half-grown state, at prices equal to early Beets sold in the same manner. Sold in this state, they are highly profitable at the prices received, but only limited quantities can be disposed of. In the dry state, during fall and winter, they range from $1.50 to $2.50 per barrel, according to quality, and at these prices will yield double the profit of Potatoes, as a farm crop.

The varieties in general cultivation are limited. The favorite variety for all purposes is the

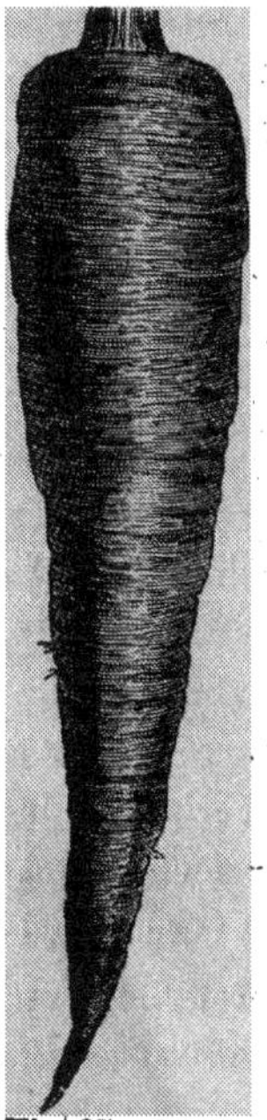

Long Orange.—This is equally adapted for garden or farm culture; it is of large size, fair specimens averaging 12 inches in length, with a diameter of 3 inches at the top; color orange-red, varying in depth of shade in different soils.

Early Horn.—An old and favorite sort for an early crop, but not large enough to be suitable for general culture. It is the variety that is sold in our markets bunched up in the green state. It matures its root eight or ten days earlier than the preceding. It is also more tender, and is more valued than any other for culinary use. It may be grown closer than the Long Orange, as its foliage is much shorter.

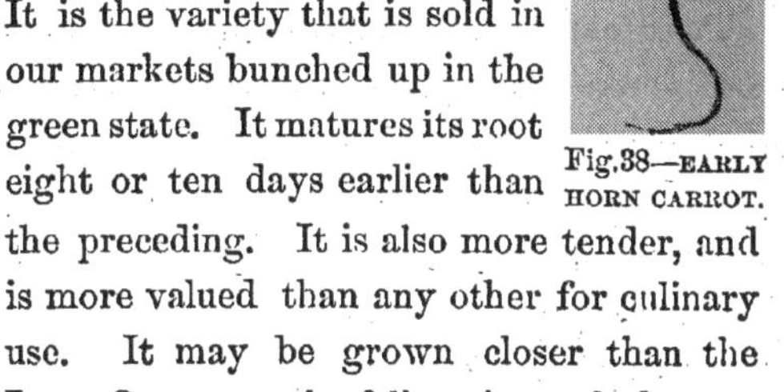

Fig. 37.—LONG ORANGE CARROT.

Fig. 38—EARLY HORN CARROT.

Early French Horn.—This variety is used only for forcing in hot-beds or vegetable forcing house, its small foliage and short root, not unlike the Turnip Radish in shape, rendering it especially suitable for growing under glass. It is not yet very generally grown for market, thus forced, but what few have been grown, were quickly sold at most profitable rates, $12 for the products of a 3×6 sash, or about 5 cents a piece.

White Belgian.—This is the most productive of all known varieties; the lower part of the root is white, that growing above the ground, and exposed to the air, green.

It is exclusively grown for stock, bearing nearly twice as much weight per acre as the Long Orange. Horses do not eat it quite so readily, however, and it is said to be less nutritious than the Red or Orange sorts.

CHERVIL.—TURNIP-ROOTED.—(*Chærophyllum bulbosum.*)

A vegetable of recent introduction, closely allied to the Parsnip, which it resembles in shape. It is of a grayish color; the flesh is white and mealy, tasting something like the Sweet Potato. It is equally as hardy as the Parsnip, and in France, where it has been cultivated to a considerable extent, is said to have yielded 6 tons per acre. It is one of the many plants that were experimented with in Europe as a substitute for the Potato, when it was feared that that root would be lost to us by disease. Its culture is in all repects similar to the Parsnip or Carrot; it is entirely hardy in any latitude, and is rather improved by the action of the frost. It must be sown as early in spring as the soil is fit to work, it being slow to germinate if the weather becomes hot and dry.

CELERY.—(*Apium graveolens.*)

I know of no vegetable on the cultivation of which there is so much useless labor expended with such unsatisfactory results, as Celery. Almost all private cultivators still think it necessary to dig out trenches, from 6 to 12 inches deep, involving great labor and expense, and giving a very inferior crop to that planted on the level surface, in the manner practised on hundreds of acres by the market gardeners in the vicinity of New York.

Our manner of treating the Celery crop, of late years, is very much simplified. Instead of sowing the seed in a hot-bed or cold frame, as formerly, it is sown in the open ground as soon as the ground is fit to work in spring—here about first week in April—on a level piece of rich mellow soil, that has been specially prepared by thorough pulverizing and mixing with short stable manure. The bed being fined down by raking so that it is clear of stones and all inequalities, lines are drawn out by the "marker" 8 or 9 inches apart, in beds of 8 rows in a bed, rubbing out every 9th line for an alley, on which to walk when weeding, etc. The seed should be sown rather thinly, one ounce being sufficient for every 20 feet in length of such a bed. After sowing, the bed should be rolled, or patted down with a spade, *which will give the seed sufficient covering.*

As soon as the seeds of Celery begin to germinate, so that the rows can be traced, hoe lightly between the rows, and begin to pull out the weeds as soon as they can be seen. One day's work, at the proper time, will be better than a dozen after the seed bed gets enveloped with weeds, besides ensuring much finer plants.

As the plants advance in growth, the tops are shorn off, generally twice before the time of setting out, so as to induce a stocky growth; plants, thus treated, suffer less on being transplanted.

Celery may be planted any time from middle of June to middle of August; but the time we most prefer is during July, as there is but little gained by attempting it early. In fact, I have often seen plants raised in hot-beds and planted out in June, far surpassed both in size and quality by those raised in the open ground and planted a month later. Celery is a plant requiring a cool moist atmosphere, and it is nonsense to attempt to grow it early, in our hot and dry climate; and even when grown, it is not a vegetable that is ever very palatable until cool weather. This our market experience well proves, for although we always have a few bunches exposed for sale in August and September, there is not one root sold then for a thousand that are sold in October and November. Celery is always grown as a "second crop" by us, that is, it follows after the spring crop of Beets, Onions, Cabbage, Cauliflower, or Peas, which are cleared off and marketed, at latest, by the middle of July; the ground is then thoroughly plowed and harrowed. No additional manure is used, as enough remains in the ground, from the heavy coat it has received in the spring, to carry through the crop of Celery. After the ground has been nicely prepared, lines are struck out on the level surface, 3 feet apart, and the plants set 6 inches apart in the rows. If the weather is dry at the time of planting, great care should be taken that the roots are properly "firmed." Our custom is, to turn back on the row, and press by the side of each plant gently with the foot. This compacts the soil and partially excludes the air from the root until new rootlets are formed, which will usually be in forty-eight hours, after which all danger is over. This practice of pressing the soil closely around the roots is essential in planting of all kinds, and millions of plants are annually destroyed by its omission. After the planting of the Celery is completed, nothing further is to be done for six or seven weeks, except running through between the rows with the cultivator or hoe, and freeing the plants of weeds until they get strong enough to crowd them down. This will bring us to about the middle of August, by which

Fig. 39.—CELERY AFTER "HANDLING."

time we usually have that moist and cool atmosphere essential to the growth of Celery. Then we begin the "earthing up," necessary for blanching or whitening that which is wanted for use during the months of September, October, and November. The first operation is that of "handling," as we term it, that is, after the soil has been drawn up against the plant with the hoe, it is further drawn close around each plant by the hand, firm enough to keep the leaves in an upright position and prevent them from spreading, which will leave them as shown in fig. 39.

This being done, more soil is drawn against the row, (either by the plow or hoe, as circumstances require), so as to keep the plant in this upright position. The blanching process must, however, be finished by the spade, which is done by digging the soil from between the rows and banking it up clear to the top on each side of the row of Celery, as in fig. 40. Three feet is ample distance between the dwarf varieties, but when "Seymour's Superb,"

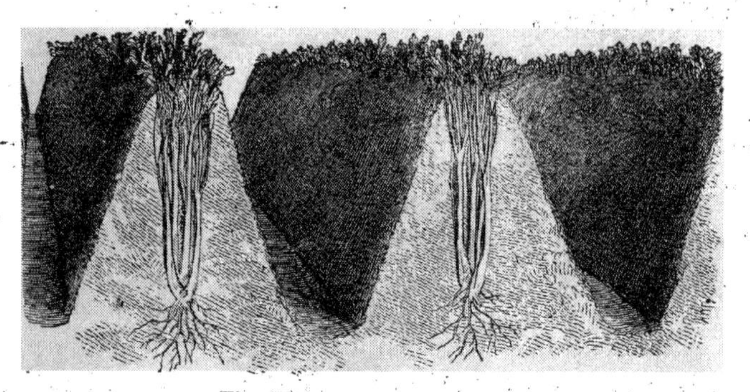

Fig. 40.—CELERY EARTHED UP.

"Giant," or other large sorts are used, the width between the rows must be at least 4½ or 5 feet, which entails much more labor and loss of ground. For the past eight years I have grown none but the dwarf varieties, and have saved in consequence at least one-half in labor, and one-third in ground, while the average price per root in market has been always equal and occasionally higher than for the tall growing sorts.

My neighbors around me have at last got their eyes opened to the value of the dwarf sorts, and I think that a few years more will suffice to throw the large and coarse-flavored sorts, such as "Seymour's Superb," and "Giant," out of our markets.

The preparation of the soil and planting of Celery for *winter use*, is the same in all repects, except that, what is intended for winter need never be "banked up" with the spade. It merely requires to be put through the handling process, to put it in a compact and upright position preparatory to being stowed away in winter quarters. This should not be done before the middle of September, or, just long enough before the Celery is dug up, to keep it in the upright position.

We have, however, another method which we have found to answer very well for the late crop, and it is one by which more roots can be grown on the same space and with less labor than by any other. It is simply to plant the Celery 1 foot apart, *each way*, nothing farther being required after planting, except twice or thrice hoeing to clear the crop of weeds until it grows enough to cover the ground. No handling or earthing up is required by this method, for, as the plants struggle for light, they naturally assume an upright position, the leaves all assuming the perpendicular instead of the horizontal, which is the condition essential before being put in winter quarters. This method is not quite so general with us as planting in rows, and it is perhaps better adapted for private gardens than for market; as the plant is more excluded from the air, the root hardly attains as much thickness as by the other plan.

Our manner of preserving it during winter is now very simple, but as the knowledge of the process is yet quite local, being confined almost exclusively to the Jersey market gardeners, I will endeavor to put it plain enough, so that my readers "may go and do likewise." In this locality we begin to dig up that which we intend for winter use about the end of October, and continue the work (always on dry days) until the 20th or 25th of November which is as late as we dare risk it out for fear of frost Let it be understood that Celery will stand quite a sharp frost, say 10 or even 15 degrees, while 20 or 25 degrees will destroy it. Hence experience has taught us, that the sharp frosts that we usually have during the early part of November, rarely hurt it, though often causing it to droop flat on the ground, until thawed out by the sun. It must, however, never be touched when in the frozen state, or it is almost certain to decay. The ground in which it is placed for winter use should be as dry as possible, or if not dry, so arranged that no water will remain in the trench. The trench should be dug as narrow as possible, not more than 10 or 12 inches wide, and of the depth exactly of the hight of the Celery; that is, if the plant of the Celery be 2 feet in length, the depth of the drain or trench should be 2 feet also. The Celery is now placed in the trench as near perpendicular as possible, so as to fill it up entirely, its green tops being on a level with the top of the trench. Figure 41 represents a section across a trench filled with Celery in the manner just described. No earth whatever is put to the roots other than what may adhere to them after being dug up. It being closely packed together, there is moisture enough always at the bottom of the trench to keep this plant, at the cool season of the year, from wilting. That which is put in trenches about the 25th of October, is usually ready to be taken up for use about the 1st of December, that a couple of weeks later, by 1st January, and the last (which we try always to defer to 15th or 20th November) may be used during the winter and until the 1st of April. For the first lot, no covering is required, but that for use during the winter months, must be gradually covered up, from the middle of December, on until 1st of January, when it will require at least a foot of covering of some light, dry material—hay, straw, or leaves—the latter perhaps the best

Fig. 41.—CELERY STORED FOR WINTER.

I have said the covering up should be gradual. This is very important, for if the full weight of covering is put on at once, it prevents the passing off of the heat generated by the closely packed mass of Celery, and in consequence it to some extent "heats," and decay takes place. Covered up in this manner, it can be got out with ease, during the coldest weather in winter, and with perfect safety. These dates of operations, like all others named throughout, are for this latitude; the cultivator must use his judgment carefully in this matter, to suit the section in which he is located.

Regarding the *profits* of this crop I can speak from a very extensive experience in its culture, having cultivated an average of ten acres for the past eighteen years. For many years, in the early part of that time, it was, by no means, what we would now call a profitable crop. By persisting in raising the large growing sorts, and the awkward and expensive mode we had then of working it, we were satisfied if it gave us a profit of $50 or $75 per acre. But for the last six or eight years, by adopting the flat culture, and the drain or trench system for winter storage, it has done much better, and is now a very profitable "second crop," averaging a clear profit of $300 per acre, though it rarely brings over $3 per 100 roots. No doubt, in many parts of the country, it would be much more profitable than in the crowded markets of New York. It is shipped from here in all directions; to Philadelphia (largely), Baltimore and Washington, (South), and to Newport, Providence, Hartford and New Haven, (East). It is a bulky and expensive article to ship, and the dealer must realize more than double on the purchase, or it will not pay his risk. It must thus cost the consumer, in these towns to which we send it, 8 or 10 cents a head, a price at which it would pay a clear profit of $1000, or $1500 per acre.

If the awkward and laborious systems of cultivation still persisted in for the growing of Celery, is a mistake, the continued use of the tall growing and coarse varieties, we believe to be even a far greater one. The kinds that should be grown, either for private or market use, are very limited.

Incomparable Dwarf.—This, so far, is decidedly our

best variety; under good cultivation it attains a hight of 2 feet, and a circumference of 12 inches; it is perfectly solid, the stalks half round, the leaves and stems being rather light green. When blanched, it is a yellowish-white, crisp, tender, and of a most agreeable nutty flavor. The great advantage of this, and other dwarf sorts, over the large kinds, is, that nearly every part of the plant is fit to eat when blanched; for instance if in the dwarf varieties the length is only 2 feet, and in large sorts 3 feet, the extra length of the large sort is unfit for use, being usually only an elongation of the outer leaves, the *heart* or edible part rarely rising more than 18 inches in the large sorts, while the dwarf sorts may be said to be all heart. This variety, for *fall use*, is planted 3 feet between the rows, by 6 inches between the plants, or nearly 27,000 roots per acre. For *winter use*, when it does not require to be "banked," we plant 2 feet between rows, and 6 inches between plants, or about 40,000 roots per acre.

Boston Market.—A variety very similar to the above, but rather more robust, though a dwarf variety; the leaves are darker green, the stalks when blanched nearly white; it is solid, crisp, and tender; an excellent variety.

Dwarf Red.—A variety similar in all respects to the "Incomparable Dwarf," except in color of the stalks, which are of a rosy crimson; although the flavor of the red varieties of Celery is acknowledged to be superior to the white, and the appearance, when blanched certainly far richer, yet, for some unexplained reason, they do not so readily sell in our markets. In the London markets, about equal quantities of each are sold.

Seymour's Superb.—The best of the large-growing

sorts, attaining a hight, under good culture, of 3 feet. It should never be planted closer than 4 feet between the rows, or it cannot be worked properly. For southern sections of the country, this variety is more suitable than the dwarfs, as it grows freer in a hot and dry atmosphere.

CELERIAC, OR TURNIP-ROOTED CELERY.

(*Apium graveolens. Var.*)

Fig. 42.—CELERIAC.

Is grown from seeds sown in the same manner, and planted out at the same seasons as directed for Celery; but as it requires but a slight earthing up, it is planted closer than ordinary Celery, 18 inches between the rows and 6 inches between the plants. It is preserved for winter use in shallow trenches, and covered up, as the season advances, as directed for Celery. It is as yet grown to but a limited extent here, being used only by the French and Germans. The Turnip-like root is cooked, or it is sliced and used with vinegar, making a most excellent salad.

CHIVES.—(*Allium Schœnoprasum.*)

A small bulbous-rooted plant of the Onion tribe, entirely hardy and of the easiest culture, as it will grow on almost any soil for many years without renewal. It is propagated by division of the root, and may be planted at 9 or 10 inches apart; the leaves are the parts used, which may be repeatedly shorn off during the early summer months. They are sometimes used in soups, but more generally in the raw state.

CORN SALAD, OR FETTICUS.—(*Fedia olitoria.*)

A vegetable used as a salad, and sold to a consider able extent in our markets. It is sown on the first opening of spring, in rows 1 foot apart, and is fit for use in six or eight weeks from time of sowing. If wanted to come in early in spring, it is sown in September, covered up with straw or hay, as soon as cold weather sets in, and is wintered over exactly as Spinach. The covering is removed in March or April when it starts to grow, and is one of our first green vegetables in spring.

CRESS, OR PEPPER GRASS.—(*Lepidium sativum.*)

Another early spring vegetable, used as a salad, and of easy culture. It is sown in early spring in rows 1 foot apart; as it runs quickly to seed, succession sowings should be made every eight or ten days. There are sev eral varieties, but the kind in general use is the Curled, which answers the purpose of garnishing as well as for salads.

CRESS—WATER.—(*Nasturtium officinale.*)

This is a well-known hardy perennial aquatic plant, growing abundantly along the margins of running streams, ditches, and ponds, and sold in immense quantities in our markets in spring. Where it does not grow naturally, it is easily introduced by planting along the margins of ponds or streams, where it quickly increases, both by spreading of the root and by seeding. Many a farmer, in the vicinity of New York, realizes more profit from the Water Cresses, cut from the margin of a brook running through his farm, in two or three weeks in spring, than from his whole year's hard labor in growing Corn, Hay, or Potatoes.

It is usually sold in baskets containing about 3 quarts, which sell, when first in market, at $1 each; 200 or 300 such are carried in an ordinary wagon, so that from a single load of this simple vegetable, $200 to $300 are realized. The Water Cress has a particularly pleasant pungent taste, agreeable to most people in early spring.

It is said, that when Sir Joseph Banks first arrived in England after his voyage around the world, among the first things he asked for were Water Cresses, well knowing their value as a purifier of the blood; and that he afterwards presented one of the largest Water Cress growers for the London market a Banksian Medal, for energy shown in the business, believing that while he had benefited himself, he had benefited the community. I have no doubt whatever, that, in situations where irrigation could be used at pleasure, and regular plantations made as for Cranberries, that, grown in this way,—judging from the enormous price they sell at, picked up as they are in the present hap-hazard way—at present prices, an acre would sell for $4000 or $5000.

COLEWORT, OR COLLARDS.—(*Brassica oleracea.*)

Collards, as grown in this country, are nothing more than sowings of any early variety of Cabbage, in rows about one foot apart, which are cut off for use when 6 or 8 inches high. Spring sowings may be made every two weeks from April to June; and in fall from September, as late as the season will admit. I have never seen them sold in our markets.

CORN.—(*Zea Mays.*)

The varieties known as "Sweet," are the sorts most cultivated for culinary use in the green state. It may be either sown in rows 4½ feet apart, and the seeds planted at 8 or 9 inches in the rows, or planted in hills at distances of 3 or 4 feet each way, according to the variety grown or the richness of the soil in which it is planted. The taller the variety, or richer the soil, the greater should be the distance apart. The soil best suited for Corn, for an early crop, is a well enriched sandy loam. The planting should never be done until the weather is settled and warm, as heat is indispensable to the healthy growth of Corn. We make our first plantings, in this vicinity, about the middle of May, and continue succession plantings every two or three weeks until the first week in July, which date is the latest at which we can plant and be sure of a crop of "roasting ears." In more southerly latitudes, planting is begun a month earlier, and continued a month later. The crop is not profitable enough for the market garden, but the farmers realize double the price for Sweet Corn when sold in the green state in our markets, that they do for ripe Corn, besides, as the ground can be cleared when thus sold in August, it can be used afterwards for Turnips as a second crop. A profit of from $50 to $100 is said to be realized per acre from Sweet Corn.

The most popular variety is:

Early Darling, which is early, of good size, and under good cultivation, gives an average of three ears on each stalk. It is tender and sweet; as this sort is rather dwarf growing, it need never be planted more than 3 feet apart.

Dwarf Prolific Sugar.—This variety rarely grows more than 4 or 5 feet in hight, suckering up from the main stem, often five or six shoots,—which bear an average of two ears each; these, however, are small, not more than 5 or 6 inches in length, and quite slender. It is too small for market purposes, but is the most valuable variety grown for family use, being early, tender, sweet, and productive. We prefer to grow this variety in rows 3 feet apart, and 1 foot between the plants.

Asylum Sugar.—A large late variety of tall growth, used to succeed the earlier sorts; it is productive, and has every desirable quality except earliness. Should be planted in hills 4 feet by 3.

Stowell's Evergreen.—Also a late variety, having the peculiarity of remaining longer in the green state than any other sorts; on this account it is very popular.

CUCUMBER.—(*Cucumis sativus.*)

The growing of the Cucumber out-of-doors is, in most places, attended with a great deal of annoyance and loss, occasioned by the attacks of the "Striped Bug." When the seed is sown in the open ground, repeated sowings are often utterly destroyed by this pest, despite of all remedies. To avoid this, and at the same time to forward the crop at least a week, we have long adopted the following method, with the greatest success. About the middle of May, (for this section), we cut from a pasture lot, sods from 2 to 3 inches thick, these are placed with the grassy side down, either on the benches of our forcing house, in an exhausted hot-bed, or inside of a cold frame; at that season of the year any one of these will do as well as another. The sods being fitted together neatly so that all crevices are filled up; they are then cut into squares about 3 or 4 inches in length and breadth; on each of these are planted 2 or 3 seeds of Cucumber, and over the whole is sifted about half an inch of covering of some light rich mold. They are then sprinkled thoroughly from a Rose Watering-pot, and the sashes put on, and kept close until the seeds begin to germinate, which will be in three or four days. As soon as they are up, the sashes must be raised to admit air, else the sun's rays, acting on the glass, would raise the temperature too high; at that season of the year the sashes, as a rule, may be tilted up at 8 or 9 o'clock in the morning, and shut down by 3 or 4 o'clock in the afternoon. By the time the Cucumber plants have attained two or three of their rough leaves, which will be in about three weeks from the time of sowing, they are planted out in the open ground in hills 3 feet apart each way. The hills should have been previously prepared, by mixing thoroughly with the soil in each, a shovelful of well-rotted manure.

It is always better to plant in the afternoon, rather than during the early part of the day, as the coolness and moisture at night enable the plants to recuperate from the effects of removal. If the weather is hot and dry, it is safer to give each hill a thorough watering *once*, immediately after planting. I have recommended sods in preference to flower-pots for starting the Cucumbers, inasmuch as they are not only procurable in all places, but our experience is, that the sod is even better than the flower-pot; it better retains moisture, and there is a freshness about sod in which the roots of all plants love to revel, and which no composts we can prepare can ever equal. It will be seen that the expense of growing Cucumbers, in this manner, is considerable; to grow enough for an acre—about 5000 hills—it will require the use of at least 20, 3 × 6 sashes, and the preparation of the sod, and attention in airing, etc., until they are fit to plant, will involve ten times more expense than simply sowing the seed in the hills; but all such expenditures are well returned, for it is safe to say, that the profits would always be at least three times more by this plan than by the other. The average receipts are $750 per acre; working expenses probably $250, and the crop is off in time for Turnips or Spinach as a second crop.

The Cucumber is a vegetable perhaps better fitted than any other for southern market gardeners. There is no doubt, that by the forwarding process above described, it could be got in marketable condition in the neighborhood of Charleston or Savannah, at least a month before it could in New York, and as it is one of the easiest things we have to ship, a profitable business could be made of growing it to send North. The profits on an acre of Cucumbers, grown by this method in Charleston, and sold in New York in June, would, I think, exceed the average profits of fifty acres of Cotton.

Cucumbers are also extensively grown for pickling; hundreds of acres being used for this purpose in the vicinity of New York, especially in Westchester County. Sod or stubble land, plowed in early fall, and again turned over twice or thrice in spring, is the condition of soil usually chosen. The ground is marked out as for Corn, 4 feet each way, and a good shovelful of well-rotted manure, dug in at the angle which forms the hill; the seed is sown—about a dozen in each hill—usually about the 20th of June, but equally good crops can be obtained by sowings made as late as the middle of July. The average price of late years has been $1.50 per 1000, and the number grown per acre on properly cultivated lands, is 150,000, which is $225 gross receipts per acre. The expense of raising are said to be about one-half. These profits would not satisfy the market gardener on his few valuable acres near the city, but, no doubt, are remunerative enough to the farmers, with large quantities of cheap land.

The varieties are numerous, and embrace many very well marked kinds. The large growing kinds that attain 2 feet or more in length, have never become favorites in our American markets.

White Spined.—Belongs to the short growing section, is of medium size, from 6 to 8 inches in length and 2 to 3 inches in diameter; it is a very handsome variety, deep green, flesh crisp, and of fine flavor. The variety almost exclusively grown for market in New York.

Fig. 43.—WHITE-SPINED CUCUMBER.

Early Frame.—A very handsome small growing variety, rarely exceeding 5 inches in length, and has fewer spines than the preceding. It is often a question whether this or the White Spined is most desirable, so that of late years we have grown about an equal quantity of each for forcing or forwarding under glass.

Gherkin.—This variety, used exclusively for pickling, botanists distinguish as a species distinct from the common Cucumber; it is very small, length from 2 to 3 inches; a strong growing sort, and should be planted 5 feet apart.

Manchester Prize.—An extremely handsome variety, extensively grown in England; it is very dark green, having tubercular excrescences at the base of the spines, regularly over its whole surface, except 3 or 4 inches at the extremities, which are smooth; it is of the largest size, growing upwards of 2 feet in length.

Long Green Turkey.—This is a very distinct sort, slightly curved at the stem, measuring, when full grown, 15 or 16 inches; it is perhaps the firmest and best flavored of all Cucumbers, and as it has but few seeds, can be used older than most others.

Early Cluster.—A much esteemed early variety, growing in clusters and extremely productive; its color is blueish-green, shading lighter at the extremities.

EGG PLANT.—(*Solanum Melongena.*)

The cultivation of the Egg Plant, from its extreme tenderness, is, in its early stage, attended perhaps with more trouble than any vegetable of our gardens. A native of Tropical America, it at all times requires a high temperature; for this reason, in this latitude, the seeds had better not be sown in the hot-bed until first week in April, and even then a steady bottom heat is necessary to a healthy development, and there should be warm covering at night over the sashes. I have always found that in tender plants of this kind, there was nothing gained by starting early, even though by great care the plants are carried through the cold season. By the time they can be planted in the open ground, about June first, those started first of March, would be no larger than those started first of April, besides being harder both in roots and leaves, in which condition they are far inferior to the younger plants that have been raised with less than half the labor.

The soil in which Egg Plants are to be grown can hardly be too rich, for it is a plant that will generally re pay good treatment. They are planted from 2 to 3 feet apart, according to the degree of richness of soil; in the fertile market gardens never less than 3 feet. Although their sale is comparatively limited, yet from the difficulties often experienced in raising the plants, all that are offered, are sold at good prices; the average is about $1 per dozen, each plant producing six to nine full-sized fruit. It is more important with this vegetable to select the proper variety for growing, than with any other that I know; for that reason we are chary of touching any other sort for market purposes than the

New York Improved.—This is readily distinguishable from either the "Large Round" or "Long Purple" varieties, in the plant being more robust in all its parts, the leaves and stems also being thickly studded with spines, which are not to any extent on the other varieties, but the great merit it has over the others is its uniform productiveness. I grew from 1000 to 3000 Egg Plants for market for over a dozen of years, but never had a paying crop with any other sort except the *New York Improved.*

Fig. 44.—N. Y. IMPROVED EGG PLANT.

Fig. 45.—LONG EGG PLANT.

Long Purple.—Different in shape from the foregoing; sometimes deep purple, and again pale, with white or yellowish stripes.

It is claimed that this is earlier than the preceding, but this we are not willing to concede, for all our experience with them, to the best of my recollection was, that neither this or any other variety than the New York Improved, ever proved worthy of cultivation, in our vicinity at least.

Scarlet-fruited Egg Plant.—This is more grown as a curious ornamental plant than for culinary use, the fruit is about the size and shape of a duck egg, of beautiful scarlet.

White-fruited Egg Plant.—Similar in growth to the scarlet, but the fruit is larger, and of an ivory whiteness. It is good when cooked, but much less productive, and like the Scarlet, is grown more for curiosity and ornament.

ENDIVE.—(*Cichorium Endivia.*)

The cultivation of this vegetable for market purposes is not yet extensive, it being used by few except our German and French population. It is, however, offered now by the wagon load, where a few years ago a few basketfuls would have supplied all the demand. Like all other vegetables that are grown in limited quantities, it commands a high price, and the few who do raise it find it very profitable.

Like Lettuce, it may be sown at any time from early spring until August, and perfect its crop the season of sowing. As it is used almost exclusively in the fall months, the main sowings are made in June and July, from which plantations are formed, at 1 foot part each way, in August and September. It requires no special soil or manure, and after planting, it is kept clear of weeds by hoeing and weeding, until the plant has attained its full size, when the process of blanching begins; for it is never used except when blanched, as it is harsh and bitter in the green state. Blanching is effected by gathering up the leaves, and tying them up by their tips in a conical form, with bass matting. This excludes the light and air from the inner leaves, which in the course of three to six weeks, according to the temperature at the time, become blanched.

Another method is much simpler and quicker, and is the one mostly practised by those who grow Endive for market; it consists simply in covering up the plants as they grow, with slates or boards, which serves the same purpose, by excluding the light, as the tying up. The average price, during the months of October, November, and December, is $1 per dozen.

The best sorts are the following:

Green Curled.—This is not only one of the most useful as a salad, but is highly ornamental from its delicately cut and curled leaves; it is much used for garnishing.

Moss Curled.—This variety is as yet scarce, but no doubt it will soon be extensively cultivated. From the density of the foliage, the plant is heavier than the Green Curled, is equally agreeable as a salad, and its appearance, either green or blanched, is particularly handsome.

Broad-leaved Batavian. — A loose growing variety, forming but little heart. As with this blanching can only be accomplished by tying up, it is not so desirable as either of the preceding.

White Curled.—This, as the name indicates, has white or light foliage; it is more tender than the Green; it cannot be recommended except as an ornamental variety.

GARLIC.—(*Allium sativum.*)

Another vegetable used mostly by foreigners. It is of the easiest culture, growing freely on any soil suitable for Onions. It is propagated by divisions of the bulb, called "cloves," or "sets." These are planted in early spring, in rows, 1 foot apart, and from 4 to 6 inches between the plants in the rows. The crop matures in August, when it is harvested like the Onion. It is always sold in the dry state.

HORSERADISH.—(*Nasturtium Armoracia.*)

This root is now one of the most important we raise in our market gardens, upwards of two hundred acres of it being grown in the vicinity of New York alone, and for the last half dozen years there has been nothing grown from which we have realized more profit as a second crop. It is always grown as a second crop in the following manner:

In preparing the roots for market during winter, all the small rootlets are broken off and reserved for planting, leaving nothing but the main root, which is usually from

12 to 15 inches long, and weighing about three-quarters of a pound. The rootlets, or sets, are cut into pieces of from 4 to 6 inches in length, from one-quarter to one-half in diameter; these are tied in bundles of 50 or 60, the top end being cut square and the bottom end slanting, (fig. 45), so that in planting there will be no danger of setting the root upside down; for although it would grow, if planted thus, it would not make a handsome root.

Fig. 46.—HORSERADISH SET.

The sets, when prepared, are stowed away in boxes of sand, care being taken that a sufficiency of sand is put between each layer of bundles to prevent their heating. They may either be kept in the boxes in a cool cellar, or pitted in the open ground, as may be most convenient. We prefer the open ground, when the weather will permit. I have said that Horseradish is always cultivated as a second crop; with us, it usually succeeds our Early Cabbage, Cauliflower, or Beets. Thus, we plant Early Cabbage, lining out the ground with the one foot marker; on every alternate line are first planted Cabbages, which stand, when planted, at 2 feet between the rows, and 16 or 18 inches between the plants. We always finish our entire planting before we put in the Horseradish, which delays it generally to about 1st of May; it is then planted between he rows of Cabbage, and at about the same distance as the Cabbage is in the rows, giving about 12,000 or 13,000 plants per acre. The planting is performed by making a hole, about 8 or 10 inches deep, with a long planting stick or light crowbar, into which is dropped the Horseradish set, so that its top will be 2 or 3 inches under the surface; if the sets should be longer, the hole should be made proportionally deep, so that the top of the set be not nearer the surface than 2 or 3 inches; the earth is pressed in alongside the set, so as to fill up the hole as in ordinary planting. The main reason for planting the set so far under the surface, is, to delay its coming up until the crop of Cabbage be cleared off; the Horseradish makes its main growth in the fall, so that it is no injury to it to keep it from growing until July; in fact it often happens that by being planted too near the surface, or too early, it starts to grow so as to interfere with the Cabbage crop; in such cases we have often to cut the tops off twice by the hoe, before the Cabbage is ready, but this does not injure it in the least. It is a crop with which there is very little labor during summer; after the Cabbage has been cut off, the Horseradish is allowed to grow at will, and as it quickly covers the ground, one good deep hoeing is all that is required after digging out the Cabbage stumps. When grown between Early Beets, the culture is, in all respects, the same, only it is more profitable to have the rows of Beets only 18 inches apart; this of course throws the Horseradish nearer, so that when planted between Beets, it should be planted at the distance of 2 feet between the plants in the rows.

As it is an entirely hardy plant, it is one of the last roots we dig up in fall, it being usually delayed until De cember. After digging, the small roots are usually broken off in the field and stowed away in boxes, so that they can be trimmed under cover at leisure. The main root is then put away in the pits, as recommended in Winter Preservation of Vegetables, so that it can be got at as re quired during winter. The preparation for market is very simple, being merely to cut off the green tops and small rootlets, leaving the main root only, as represented, in reduced size, by fig. 47. It is sold by weight, and is generally washed, which is done sufficiently by rinsing a quantity of it together in a large tub.

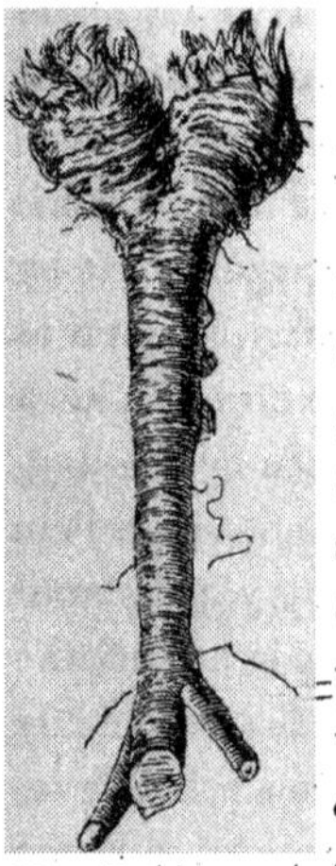

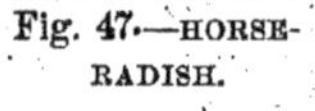

Fig. 47.—HORSE-RADISH.

Our manner of growing Horseradish in this district, we claim to be a great advance on the methods practised in general. All American writers on the subject, that I have seen, follow in the same track and recommend planting the *crowns.* This would not only destroy the most salable part of the root, but when planted thus, the crowns will produce only a sprawling lot of rootlets, that are utterly unsalable in the market.

They also tell us, that "after two seasons growth the roots will be fit for use." Now, my experience in growing this root, has most emphatically told me that after two seasons growth it is entirely *unfit* for use, or at least unfit for sale, which I suppose is about the same thing. A few years ago, one of my neighbors had a patch of about two acres, which from some cause or other he had neglected to have dug until late in spring, and concluded, as it was then rather late to sell it, he would leave it to grow over until next season. On commencing to dig it the next fall, he found that the main root, instead of being solid, as it is at one year old, had become partially hollow, and of a woody, stringy nature, and when offered to manufacturers, it was refused at any price. So there was no help for it, but dig up and throw away his entirely worthless crop at a most unusual expense, as its two year's growth had massed the whole soil with roots. This experiment of my neighbor was a loss to him of certainly not less than $1500.

Grown in the deep rich soil of our market gardens, Horseradish has been for the past twenty years one of our most profitable second crops, and as an encouragement to beginners, I will state that the price has, in this, as with most other vegetables, steadily advanced, showing that, even with increased competition, there has been more than correspondingly increased consumption. The price for five years, ending 1854, did not average more than $70 per ton; from that time to 1860 about $120 per ton; and from 1860 to 1866 fully $200 per ton. Of course the prices these later years were inflated, yet still the proportion is higher for this than for any other vegetable. Our average weight per acre is five tons, or a little over three-quarters of a pound per root for 12,000 planted. It has always been a surprise to me how the price has kept up, in view of its easy and safe culture. But there is one thing to be remembered; these heavy crops are only obtained in our gardens that are in the highest state of culture, no ordinary farm land, the first season, manure it as you might, will produce such results.

KOHLRABI, OR TURNIP-ROOTED CABBAGE.

(*Brassica oleracea. Var.*)

In general appearance, this vegetable more resembles a Ruta Baga Turnip than a Cabbage, though it is more generally classed with the latter. It is best cultivated by sowing the seeds in rows in May, June, or July, according to latitude. In this district we sow throughout June, for succession, in rows 18 inches apart, thinning out to about 8 or 10 inches between the plants. It is rather difficult to transplant, and we generally prefer to sow the whole crop from seed, and thin it out where it stands; although when the weather is suitable, the thinnings may be planted at the distances above named. It is sold in our markets in fall in the green state, in bunches containing three roots, at an average price of $1 per dozen bunches. As it is not in general use, its sale is limited. The varieties mostly cultivated are

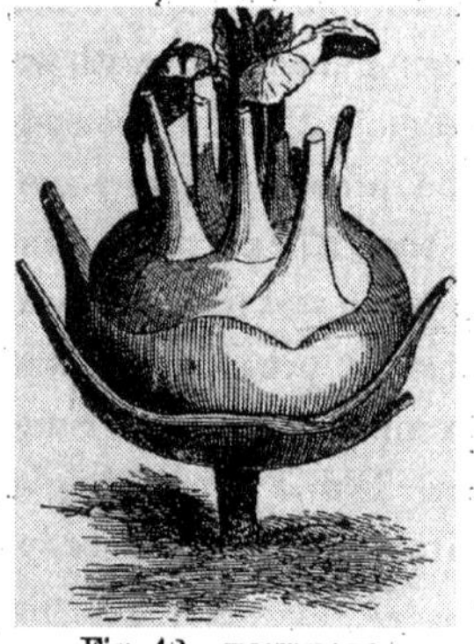

Fig. 43.—KOHLRABI.

Early White.—The bulbs are greenish-white outside; flesh white and tender, while young. The best condition for use is when the root is from 3 to 4 inches in diameter; if younger, it partakes too much of the taste of the Cabbage, and when older it is dry and stringy. The best market sort.

Large Purple.—Almost identical with the preceding, except in color, which is a blueish-purple.

LEEK.—(*Allium Porrum.*)

The Leek is another vegetable that is exclusively grown as a second crop. The seed is sown in April in rows 1 foot apart, in ground well prepared as recommended for the Celery seed bed; and like all seed beds, it is kept scrupulously clear of weeds. The best time of planting is the same as that for all our second crops; during July, or as soon as the first or spring crop can be cleared off. The ground can hardly be too rich for Leeks, and when time will allow, we always contrive to get in a slight additional coat of manure for this crop; the spring dressing, large as it always is, hardly being sufficient. The ground being well prepared by plowing and harrowing, lines are marked out by the marker at 1 foot apart, and the Leeks planted on each line at 5 or 6 inches apart; we do not earth up, but instead plant rather deeply. As it is a plant the foliage of which is but little spreading, great care must be taken that weeds are never allowed to get ahead, for if they do, they may soon entirely envelope the crop to its total destruction.

It is a vegetable used mostly in winter and spring, and requires to be dug up, in this vicinity, in November, as otherwise it would be injured by our severe winters, but in milder sections it is better left standing where it grew; it is quite a hardy vegetable, so that 20 or 25 degrees below freezing will not injure it. It is preserved, by the market gardeners here, in trenches, exactly as Celery it preserved; see chapter on Celery. Large quantities are sold in our northern markets, at fairly remunerative rates, although from the nature of the plant, it requires perhaps more labor than any other vegetable to prepare it for market. Figure 49 represents the Musselburgh Leek, trimmed previous to being bunched up for market. From six to eight roots are tied in each bunch, which bring in the market, upon an average, throughout the season, about 75 cents per dozen bunches. We plant about 85,000 plants on an acre.

Fig. 49.—MUSSELBURGH LEEK.

The two varieties used, are known as Musselburgh and London Flag. The former is rather preferred in market, being usually larger, but there is but little choice between them.

LETTUCE.—(*Lactuca sativa.*)

Perhaps there is no plant of the garden that we could so ill afford to dispense with as Lettuce. Its cultivation is universal, by all classes, and from its tractable nature and freedom from nearly all insects and diseases, it is manageable in the hands of every one. In a well appointed market garden, it is the most important vegetable cultivated, engaging our attention throughout the entire year, either in the open ground in summer, in forcing houses or hotbeds in winter, or in cold frames in spring. As our mode of growing it under glass is sufficiently described in treating of pits, frames, etc., I will in this place confine myself to our system of cultivating it out of doors.

For our main early crop, that is sold from the open ground in the latter part of May or first of June, the seed is sown the previous season in the open ground, from the 15th to 25th of September. These plants are usually large enough to be planted in frames in four or five weeks later, as recommended for Cabbage plants, about 600 or 700 being planted in a 3×6 sash. Occasionally we sow them in the frame in fall, and do not transplant them, as it saves a great amount of labor, but they are not quite so good plants unless transplanted, as it is difficult to sow them so that they come up at the regular degree of thickness. The winter treatment of Lettuce plants is similar in all respects as described for Cabbage plants. In dry, well sheltered spots, by covering up with leaves or litter, late in the season, say middle of December, Lettuce plants may be saved over winter without glass covering, and in southern parts of the country, without difficulty. Like most plants that we term hardy, 20 degrees of frost will not injure them. The plants for setting out in spring, are also sown in cold frames in February, and in hot-beds in March, and by careful covering up at night, make plants to set out in April; but these are never so good as those wintered over, and it can only be recommended where circumstances do not permit the other method. To economize, not only in space, but in manure, we make every foot of our gardens available, so that when we come to plant out our Lettuce in March or April, instead of planting it in a bed exclusively for itself, it is planted at the same time and between the rows of Early Cabbage or Cauliflower, which are set at 2 feet apart. The Lettuce matures its crop in half the time that it takes for the Cabbage, and is consequently all cut off and marketed before the Cabbage is half grown. If it were not so, they could not be both grown at once on the same ground, for, when the Cabbage crop attains its growth, it requires the full space allowed —2 feet—for its development.

This early crop of Lettuce from the open ground is considered rather an auxiliary than a main one, it meets with a rapid sale at an average of $1.50 per 100 heads. Rather a low price it may be thought; but growers, having only ten acres of ground, not unfrequently plant over 100,000 heads. It is planted somewhat closer than Cabbage, usually about 15,000 per acre. For succession crops of Lettuce, sowings may be made in the open ground as early as spring opens, until July. When not planted between Cabbage, they are set at 1 foot apart each way. They are sold to some extent throughout the whole summer, but the great weight of the crop is sold about the first week of June, in New York markets. The summer price of Lettuce is very variable, as the supply is often irregular—it may average $2 per 100.

The varieties of Lettuce used for the different purposes of forwarding and forcing, and for out-door culture in spring and summer, are of more importance than with most vegetables. I once lost almost my entire crop of Frame Lettuce, from planting the Curled India, a summer variety, instead of the Curled Silesia, which I had got from a well meaning but not very learned friend, whose hieroglyphics had got transposed.

Early Curled Simpson.—This variety we place at the

head of the list, believing it to be most generally useful for all purposes. It is a sub-variety of the Curled Silesia, rather more curled, and having more of a yellow shade of green; it is the kind which is planted in cold frames almost exclusively, and is also largely grown as an early open air variety between the Cabbage crop. This vari ety, properly speaking, does not head, but forms a compact close mass of leaves. This condition of maturity is attained quicker than in varieties that form firm heads, which gives it the great desideratum—earliness.

Curled Silesia.—A variety extensively sold in all seed-stores, but we believe the preceding to be an improvement upon it. The Curled Silesia is darker green, rather less curled, and does not mature quite so early as the Simpson.

Green Winter Lettuce.—This, next to the Brown Dutch, is the hardiest of all varieties, and for that reason it is always largely grown; many of the other varieties failing in our frames in winter, while this comes through unscathed. It is not much used for forcing, unless when we are obliged to resort to it by having lost the others. When fully matured, it forms a solid head.

Tennis Ball.—A favorite forcing variety, and, as the names indicates, forming a hard head; it makes few outer leaves, and for this reason can be planted quite closely under glass, from 6 to 7 inches apart. It is the variety mainly used in our hot-bed and forcing pits.

Black-seeded Butter.—A variety similar to the Tenni Ball, but larger in all its parts, forming heads in the open ground often 14 inches in diameter, and averaging fully a pound in weight. We use it extensively as an early out-door variety, and also to some extent in our cold frames.

Curled India.—This we consider to be by far the best of the curled sorts for hot weather; it forms an immense head, often 16 inches in diameter, and as solid as a Cabbage.

Ne Plus Ultra.—A new variety with plain leaves, which for the past two seasons, has proved particularly adapted for hot weather; it forms a large solid head, and is a very tender sort.

Paris Green Cos.—Although the Cos varieties of Lettuce are not so suitable for our climate, yet they are so much superior in flavor, that they are occasionally grown in early spring and fall for private use, though I have never seen them in our markets. In shape, they differ entirely from the other varieties, the head being elongated and of a conical form, 5 or 6 inches in diameter, and 8 or 9 inches in hight. The present variety is deep green. To be had in perfection, it requires to be tied up to ensure blanching.

Brown Cos.—Similar to the above, except in its outer coloring, which is brownish-green. This variety is said to have been cultivated in England for half a century, and is still one of the favorite sorts.

MARJORAM.—Sweet.—(*Origanum Marjoram*)

A well-known aromatic herb, grown to a considerabl extent for market purposes; but as it is of less importance than Thyme, which is grown very largely for sale, the reader is referred to that head for all information regarding its culture, price, etc., as in these particulars they are entirely identical.

MELON.—Musk.—(*Cucumis Melo.*)

The Melon is not cultivated in the vicinity of New York, in the market gardens proper, but to a large extent in what may be called the farm-gardens of Long Island and New Jersey. There it is grown almost in the same manner as they grow Corn, planting about the same time, and cultivating in the same manner, and often with but very little more profit per acre than a crop of Corn. But the Melon is a fruit easily shipped, and when grown by the forwarding process we detail for Cucumbers — by planting the seeds on sods under glass — there is no question but that it can be made equally profitable in all respects with the Cucumber, when grown in southern latitudes for our northern markets; although like many other fruits and vegetables so easily raised, it can never be expected to be profitable if used in the district in which it is grown. It is a plant not at all particular as to soil, provided it be not wet or heavy; moderately enriched light soils are those most congenial to it.

For open field culture, they should be planted in hills 6 feet each way, incorporating well with the soil, in each hill, a couple of shovelfuls of thoroughly rotted manure. Sow four or five seeds in each hill,* and cultivate afterwards as for Corn. Too much care cannot be used in saving the

* The term, "hill," when used here and elsewhere, means but a slight, if any, elevation of the surface, and is used only as a convenient term to denote where the seed or plant is to be planted. But from the signification of the word, it naturally leads the novice in gardening into very serious error, by causing him to think he must literally raise a hill on which to sow or plant, and in consequence we too frequently see plants elevated on little knolls a foot or more above the general surface, from which the rain slides, and on which the sun beats to their utter destruction.

seeds of all plants of this class, as they cross very readily; for this reason, when seeds are to be saved from any particular variety, care must be taken that it is widely removed from any other—not less than 100 yards. If seed is not wanted, improper crossing will make no difference to the fruit that season, but the seed—the product of that crop—will be affected thereby, and its impurity developed when fruit is raised from it.

From the facility with which the Melon intermixes, it follows that the varieties are very numerous, and for the same reason it is difficult to retain varieties pure. The following seem to be the most fixed in character, and are the sorts in general use.

Green Citron.—Fruit medium size, deeply netted; in shape almost round, varying from 6 to 8 inches in diameter; flesh green, and of rich delicious flavor. It is the leading market variety at the North. In selecting for seed or for use, the most netted specimens should always be chosen, as they are always vastly superior in flavor to those with smooth skin.

Borneo.—Rather a new variety, which promises to become very popular. It is of large size, nearly double that of the preceding, of a roundish oval shape, deeply ribbed and netted; flesh yellowish-green; of excellent flavor.

Nutmeg.—Fruit, nutmeg-shaped. Skin deep green, finely and thickly netted; flesh greenish-yellow, rich and sugary, in flavor second to none. It is extensively grown in private gardens, but is thought not to be quite so productive as the Citron for market.

White Japan.—A very distinct sort, deeply ribbed, slightly netted with skin of a yellowish white; flesh yellow, melting, and richly flavored. A recently introduced variety of great merit.

Skillman's Netted.—Form roundish oval; flesh deep green, sweet, and richly perfumed. A variety much esteemed for forcing, as it is closer growing than most other sorts, and also one of the earliest.

Persian Ispahan.—A most valuable variety for the more Southern States, although entirely too late for this vicinity, unless forwarded previously under glass before planting out. It is oval, of the largest size, averaging a foot in length, with a diameter in its greatest thickness of 8 inches. Skin smooth, and when fully ripe, of a light yellow; flesh yellowish white, having a rich perfume and delicious flavor.

Christiana.—A comparatively recent variety that has hardly come up to first expectations; it is reddish yellow fleshed, very sweet, but without the rich flavor of the green-fleshed sorts. However, its inferiority in flavor, is compensated by its greater earliness, but wherever the green-fleshed varieties mature, the red or yellow fleshed need never be grown.

MELON.—WATER.—(*Citrullus vulgaris.*)

Like the Musk Melon, when cultivated for sale, this is essentially a plant more suited for the farm than the garden, as it requires even more space in which to grow. The soil best suited to it is a rather sandy loam; heavier soils being apt to induce a too strong growth of leaves. It should be planted about the same time, and cultivated in all respects the same as the Musk Melon, only that the hills should not be less than 8 feet apart. It is grown in immense quantities on the light sandy soils of New Jersey, for the New York and Philadelphia markets, in which it is said to give a profit of from $100 to $200 per acre. But it is in the vicinity of our southern cities, Charleston, Savannah, etc., where we have regular steamboat communication, that these tropical fruits can be grown at a high rate of profit to the cultivator.

Water Melons, grown in the vicinity of Charleston, in July of this year, were sold by the thousand in New York, to the consumer, at $1 each; while those grown in southern Jersey, in August or September, were a drug at 15 cents and 25 cents. The leading varieties are as follows:

Mountain Sprout.—An old and well-known variety, and still the leading market sort. It is of the largest size, longish oval; skin dark green, marbled with lighter shades; red fleshed; of excellent quality. A greater weight can be raised per acre of this than of any other variety.

Black Spanish.—Fruit medium, almost round; skin dark green; flesh red; sweet and delicious. It is a great favorite in private gardens, and is claimed to be earlier than he preceding.

Ice Cream.—A round white-fleshed variety of good flavor, but not equal to the red-fleshed. It matures 10 days earlier, however, and on that account is worth cultivating in the Northern or Eastern States, where the season for the finer sorts is too short.

Orange.—So called from its peculiarity of the flesh separating from the rind when fully ripe; it is below medium size; flesh red; sugary, but coarse grained; inferior to many others.

Citron Water Melon.—Outwardly a very handsome fruit of small size, but not worth growing as a ripened fruit. It is used to some extent for making sweetmeats, for which purpose it is alone cultivated.

MINT.—(*Mentha viridis.*)

A hardy perennial plant, found growing in abundance along the roadsides in many places. It is often grown in gardens, however, and is used in soups, sauces, salads, etc., very generally. It is of the easiest culture. It is increased by divisions of the root, and planted at distances of a foot apart; it quickly forms a mass, which may be cut from for many years without renewal. It is grown to a con siderable extent in hot-beds and forcing pits, in the same way as Lettuce, and sold in the markets in early spring. Its treatment there is very simple, being merely to lift up the roots in solid mass, placing them on the 3 or 4 inches of earth in the hot-bed or bench of the forcing house, and water freely as soon as it begins to grow The sale is not large, but growers realize about $15 pe sash, (3×6), for what is thus grown throughout winter and spring.

MARTYNIA.—(*Martynia proboscidea.*)

A vegetable used to a considerable extent for pickling, the pods or fruit of which are produced in the greatest

Fig. 50.—MARTYNIA.

abundance. It is cultivated by sowing in open ground, in April or May, and transplanting to 2 feet each way, in June; it is fit for use in July and August.

MUSTARD.—(*Sinapis alba, and S. nigra.*)

Is used for culinary purposes as a salad, usually in conjunction with Cress. For this purpose it may be had throughout the entire season, by sowing during winter in hot-beds or forcing pits, and, on the opening of spring, in the open ground, where, by sowing at intervals of two or three weeks, it may be had in proper condition for use throughout the year. Sow thickly in rows, 1 foot apart, on any soil.

White Mustard.—This is the variety mostly used as a salad; the seeds are bright yellow, and are used in the manufacture of the mustard of commerce.

Black Mustard.—This variety is hardly distinguishable from the preceding, except in the color of its seeds, which are dark-brown; they are used for medicinal purposes, and also in manufacture of Mustard.

MUSHROOM.—(*Agaricus campestris.*)

I know of no vegetable which has such a novelty and interest to the beginner, as the cultivation of the Mushroom. In all other vegetables he sees something tangible to start with—seeds, plants, or roots; but here, we may almost say he sees neither, for the seeds cannot be seen with the naked eye, and it requires an unusual effort of the imagination, to believe the white moldy subtance we call *spawn*, to be either plants or roots. There are so many different systems of growing the Mushroom, detailed in most of the works on gardening, that the reader is too often bewildered in choosing a guide. In this, I will only detail one method, which I have practised for many years with unfailing success. To make the cultivation of the Mushroom profitable, it must be done in a building, either specially erected for the purpose, or in some shed, stable, or cellar, already on the premises, and which can be converted to that use. The most suitable place, in establishments having green-houses, vineries, or forcing pits, are the back sheds, usually erected over the boiler pits, such as are shown in the plans of forcing pits in this work. But such an erection is not indispensable; any place, where a temperature from 40° to 60° can be sustained during winter, will suit. We have also grown them under the stages of our green-houses, but our "modern improvements" of late years, allow us no longer room for the operation there. The time of beginning may be any time during winter; we have usually begun our preparations about December 1st, which brought our beds into bearing about February 1st, at the season that Mushrooms begin to be most wanted.

Fig. 51.—MUSHROOM.

Our method of growing Mushrooms is very simple, and can be accomplished to a certainty by any one conforming to the following directions. Let fresh horse droppings be procured from the stables each day, in quantity not less, perhaps, than a good barrow load; to every barrow load of droppings add about the same weight of fresh loam, from a pasture or sod land, or soil of any kind, in fact, that has not been manured; (the danger of old manured soil being, that it may contain spurious *fungi*). Let the droppings and soil be mixed together, day by day, as the droppings can be procured; if they can be had all at once, in quantity enough, so much the better. Let the heap be turned every day, so that it is not allowed to heat violently, until you have got quantity enough to form the bed of the dimensions required. Be careful that your heap is under cover, so that it cannot possibly get wet.

The most convenient size for a bed is from 4 to 5 feet in width, and if the Mushrooms are wanted in quantity, it is the plan most economical of space to start on the floor of the house with the first bed, the additional ones to be formed of shelving, 4 feet wide by 10 inches deep, raised one above another, something like the berths in a steamboat. Now, from the prepared heap of droppings and soil, spread over the bed a thin layer; pound this firmly down with a brick; then another layer, pounded down as before, and so on until it reaches a depth of 8 inches. Be careful that it be no more nor less than 8 inches; more would cause the mass to heat too violently, while less, is hardly enough. Into this bed plunge a thermometer; in a day or two the bed will heat so that it will run up to 100 or over, and as soon as it declines to 95 or 90, take a dibber or sharp stick and make holes all over the bed, at the distance of about 12 inches each way, to about half the depth of the bed; into each hole put a piece of spawn of the size of a hen's egg or so, covering up the hole again with the compost, so that it will present the same level firm surface as before the spawn was put in. Let it remain in this condition for about 10 or 12 days, by which time the spawn will have "run" through the whole bed. Now spread evenly over the surface of the bed nearly 2 inches of fresh loam, firm it down moderately with the back of a spade, and cover up the bed with 3 or 4 inches of hay or straw; this completes the whole operation of "planting the crop." Nothing now remains to be done, but attention to the proper degrees of heat and moisture.

If you can control the means of heating, so that the place can be kept uniformly at a temperature of 60°, all the better, but if not, it may range from 40° to 60°; it should never get much below 40°, else the bed will become cold too quickly, and delay the crop until too late in the season to be profitable. Unless the air of the house has been unusually dry, the Mushrooms will appear before any water is required; but examination should be made, and if the surface of the bed appears dry, a gentle sprinkling of water, heated to about 100°, must be given.

With this treatment, beginning in December, our first crop is ready for use in February, and as the Mushrooms do not come up all at once, it takes about three weeks to gather the crop. After this, a slight dressing of fresh soil, of about half an inch in depth, is spread over the bed, again beat down with the spade; this, gently watered with tepid water when dry, and a second crop of Mushrooms, (often better than the first), is gathered in March.

To show how a simple oversight in our operations may defeat the whole work, I will state, that in my first attempt at Mushroom growing, I labored for two years without being able to produce a single Mushroom. In my apprentice days, we had known no such word as fail, in so simple a matter; but here, on my first attempt, on my own responsibility, I was met by total failure. Every authority was consulted, all the various methods tried—but with no better success. In all such cases something must be blamed, and I pronounced the spawn as worthless; but my good natured employer quietly suggested that this could not well be, as a friend of his had abundant crops growing from spawn received from the same source. Driven into a corner by this information, I made another exploration of my "authorities," and was fortunate to find in one of them a single sentence that at once showed where my error had been, it was to "be careful to delay the covering with mould, until ten or twelve days after the bed had been spawned." Now, in all the different methods I had tried, I had in each invariably put in the spawn, and at once put on the 2-inch covering of soil, which had the effect to shut down the steam, thereby raising the temperature in the bed to a degree to destroy the spawn, and consequently to defeat my whole operations. My excuse for this digression is to show the importance of what might otherwise be thought unnecessary details.

Although spawn is procurable at cheap rates in all horticultural stores, yet to such as desire to make it themselves, I give the following brief directions. Take equal portions of horse droppings, cow dung and fresh loam, mix the whole thoroughly together, as you would make mortar; then form it into cakes about the size of large bricks, place these on edge, under cover, until they become half dry; then insert into each a piece of spawn half an inch or so square, let the bricks remain until they are quite dry; then spread about 8 inches of horse dung over the floor of the shed, on which build the bricks in a pile 3 feet wide by 3 feet high, keeping the side in which the spawn has been put uppermost; then cover them over with sufficient stable manure, so as to give a gentle heat through the whole. In two or three weeks the spawn will have spread itself through the whole mass of each brick; they are then removed to a dry place, and will retain their vital properties for many years.

There is not the least question that the cultivation of Mushrooms for market, forced in the manner detailed, will give a larger profit for the labor and capital invested than that from any other vegetable. The supply has never yet been half enough, and sellers have had prices almost pretty much as they pleased. I know of no house that has been especially erected for the purpose, and the markets have been supplied from beds formed in out-of-the-way corners, giving only an uncertain and irregular supply, very discouraging to buyers. I have no doubt whatever that Mushroom houses, roughly built, but exclusively devoted for that purpose, would, in the vicinity of any of our large cities, pay a profit of 50 per cent. per annum on cost of construction.

NASTURTIUM.—INDIAN CRESS.—(*Tropæolum majus.*)

A plant at once highly ornamental and useful. The shoots and flower-buds are sometimes used as a salad, but it is mainly grown for its seeds, which are pickled in vinegar, and used as a substitute for capers. It can be grown in almost any soil or situation, entwining itself around strings, brush, or trellis work, that may be given for its support. It may be sown thinly in drills, an inch or so deep, in April or May. There are many beautiful varieties, but the following are only those in use in our vegetable gardens.

Tall Nasturtium.—Attains a hight of 8 or 9 feet, the flowers are yellow, blotched with crimson. This variety makes an excellent screen or covering for unsightly places in the garden.

Dwarf Nasturtium.—This, in growth, is quite distinct from the preceding, never attaining more than 3 feet; it should be sown thinly in drills, 3 feet wide, and staked up with brush like Peas. Its flowers are very handsome, bright yellow, blotched with scarlet.

OKRA OR GUMBO.—(*Abelmoschus esculentus.*)

This vegetable is extensively grown in the Southern States; its long pods, when young, are used in soups, stews, etc., and are believed to be very nutritious. It is of the easiest culture, and grows freely, bearing abundantly on any ordinary garden soil. It is sown at the usual time of all tender vegetables—in this district in

May—in drills 2 inches deep, and from 18 to 24 inches apart in the rows, for the dwarf sorts, for the tall nearly double that width. There are only two sorts commonly

Fig. 52.—OKRA, OR GUMBO.

grown, "dwarf" and "tall;" the former being the most productive.

ONION.—(*Allium Cepa.*)

Next to Cabbages, perhaps, Onions are the most profitable crop of our market gardens, in which they are grown from sets, and nearly all sold in bunches in the green or unripened state. Grown from seed, they are cultivated almost exclusively by farmers or men who devote farm land to this purpose alone; thus grown, they are all sold in the dry state, and form an important article of commerce.

I will first describe the manner of cultivating in our market gardens. To produce the "sets," or small bulbs, that are planted to give early Onions to be sold green, a poor piece of ground is chosen as early as it is fit to work in spring. It is brought into a thorough state of pulverization by plowing, harrowing, and raking, so that the surface is level and free from stones; a line is then stretched, and lines are marked out by the 9-inch side of the marker, in these the seed is sown in beds of 6 rows wide, rubbing out every 7th row marked, so that it forms an alley 18 inches wide. For this purpose the seed is sown quite thickly, and on poor soil, so as to produce the "sets" as small as possible, for we find that whenever they much exceed half an inch in diameter, they will run to seed. It matters not how small the bulb is; even when of the size of the smallest Peas, they make an equally good if not a better crop, than if of a larger size. The sets are taken up in August, well dried, placed with the chaff among them in a loft of stable or barn, about 4 inches deep, covered up by six inches of hay on the approach of hard frost, and left thus until wanted for setting out in spring. Here we again commence our operations for the crop; this time the Onions are wanted as large as they can be got, and the best soil of the garden is chosen, manured with short, well rotted manure, plowed in at the rate of 75 tons to the acre; when only concentrated manures can be obtained, crushed bone is preferable to guano. The ground is further deeply harrowed; the harrow turned on its back, and the soil still further broken up with the short teeth, and if any inequalities are left, they are leveled and smoothed with the rake. The line is now stretched along the bed, and the 9-inch marker again makes the drills, 6 in each bed, with one rubbed out for an alley. The sets are now planted in the drills, at a distance of 3 inches apart, pressing each bulb down firmly, so that it will keep right side up; the row is then closed in by the feet or a rake, so that the set is entirely covered up. The ground is then rolled over, so as to render it still more compact around the bulbs; as soon as the lines can be traced, by the Onions starting to grow, the hoe is applied between the rows, and the soil broken between the plants by the fingers, where the hoe cannot reach, so as to destroy the germ of the weeds. If attended to in time, twice going over with hoeing and weeding is sufficient until the crop is fit for market, which it begins to be about the first week in June, and is usually all gathered by the first week in July, so as to give us time for second crops.

When we first begin to send them to market, they are usually not more than half grown, and are washed and tied in bunches containing from nine to twelve Onions; later, when full grown, from six to seven. This crop is one requiring considerable labor and expense, to get it in shape to sell, taking cost of sets, labor, manure, etc., probably not less than $400 per acre, for the past five years; but the receipts have been correspondingly high, averaging in that time quite $800 per acre. Onions, planted from sets, rarely fail to give a crop on any kind of soil, provided it has been well manured; and although they are sold by the market gardeners in the green state, they are equally good, ripened and dried, when raised from sets, as from seed. The quantity of sets required per acre, is from six to ten bushels according to size; at present prices, they cost $10 per bushel.

The method of raising Onions from seed as a farm or garden crop, differs but little from that we adopt for sets, except that they are sown generally about 1 foot apart in the rows, and manured at the rate of only 25 tons to the acre instead of 75. It is of the utmost importance that the ground for Onions, grown from seed, be as nearly level as possible, so that the seed is not washed away by rains. It also saves considerable labor in hoeing, when it can be got free from stones and seeds of weeds.

The seed of Onions, when sown as a field crop, is mostly sown by a machine used solely for that purpose; this machine sows two rows at once, making the drill and sowing as it goes along. The operation of sowing is begun as soon as the ground is fit to work in spring, as we always find, other conditions being the same, that those earliest sown produce the heaviest crop. The covering of the seed is best done by rolling the ground with a light roller, drawing it lengthwise of the lines. Rolling is also of great advantage in smoothing the surface, so that hoeing, particularly with the scuffle hoe, is much more easily performed. The quantity of seed per acre, is about 4 lbs., when sown by the machine, that is if the seed is new and fresh; and none else should ever be used, as Onion seed, of more than one year old, is not apt to produce a vigorous crop. The seed is quickly tested by placing a little of it in damp cotton or moss, in a moderately warm room; if fresh, it will sprout in three or four days. The early attention to weeding or hoeing is, if possible, of more importance when the crop is raised from seed, than when from sets; for the growth being slower and feebler from seeds, if weeds once get ahead, the crop may be ruined. Here, as well as in all other garden operations, one man will hoe over more ground, before the weeds start to grow, than ten men will, after the weeds get to be six inches high. Here then, a "stitch in time," literally, "saves nine."

The crop is always harvested in August, the bulbs being lifted by slightly digging under the row with a light digging fork. The Onions are left on the ground, usually from two to three weeks, according to the condition of the weather, to get thoroughly dried, and are then placed in barrels, or about 6 or 8 inches deep upon shelving made for the purpose, in a barn or cellar; any place that is dry, without being too warm, being most suitable. Onions will only endure a certain amount of frost without injury, so that it is always safer to cover them up from intense freezing as cold weather approaches. The price of Onions is variable in the extreme; those first sent to market often selling for $5 and $6 per barrel, while in a week later the same quality is hardly salable at $1.50 per barrel. Again, in spring, when successfully wintered over, some careful grower often realizes the first named price.

It may be given, as a fair average, that $1.50 per barrel is the price realized by the grower. The product is about 150 barrels per acre, and the cost of raising about $125; leaving a profit of $100 per acre.

The varieties of Onion are quite numerous, but, as in all other leading vegetables, cultivators confine themselves to only a few well established sorts.

Yellow Dutch, or Strasburg.—This is the variety that is grown almost exclusively when the crop is planted from "sets," and sold in the green state. Not that it is any more productive, nor does it sell quite so well as the White or Silver Skinned; but from the fact of the "sets" keeping better in winter, it is the sort we find safest to use.

White, or Silver Skinned.—A very handsome sort, of delicate flavor, much grown for private use; it requires more care in keeping in winter than any other, however, and is hardly ever used as a market sort, unless for pickling, for which it is sold in large quantities. The bulb is quite flat; the outer skin, silvery white. It is entirely distinct from the "Silver Skin" of the Eastern States, which is a brownish-yellow skinned variety, better known as Old Yellow, or Common Yellow.

Fig. 53.—SILVER-SKINNED ONION.

Yellow Danvers.—This variety has not been sufficiently tested as a market sort, to be sold green, but from what I have seen of it, I am inclined to think it may yet supersede the Strasburg for that purpose. It certainly is far more productive, but the question, whether it will keep as well in sets, during winter, has not been sufficiently test-

Fig. 54.—YELLOW DANVERS ONION.

Fig. 55.—WETHERSFIELD RED ONION.

ed. When sown from seed, as a field crop, it is said to give one-third greater weight than any other variety, but it does not keep so well as some others.

Wethersfield Large Red.—This is the staple variety of the eastern Onion growers; enormous quantities of it are grown for shipment, it being found, from its excellent keeping qualities, to be best fitted for that purpose. It is never so salable as the White or Yellow, however, in our home markets, and is rarely grown from sets here.

Fig. 56.—POTATO ONION.

Potato Onions, or "Multipliers," as they are sometimes called, are the mildest of all Onions, and though not generally grown for market, are perhaps the best of all for family use. They are grown by planting the small bulbs, early in spring, in rows 1 foot apart, by 4 or 5 inches in the row, and cultivated otherwise as described for those grown from sets. The increase is formed by the bulb, as it grows, splitting up and dividing into six or eight bulbs, these forming the crop when at maturity in August.

Top, or Tree Onion.—Has a bulb in size and general appearance similar to the Potato Onion, but is propagated by the singular production of a cluster of small bulbs in place of flowers. These resemble a cluster of hazel nuts, and by them it propagates very rapidly. It is grown to a considerable extent, in some places, as an early market sort, sold green; when ripe, it must be used early in the season, as it does not keep well in winter. The planting and subsequent culture is the same as for other sorts.

Fig. 57.—TOP ONION.

PARSLEY.—(*Petroselinum sativum.*)

A vegetable in more general use for garnishing than any other plant of our gardens; it is also extensively used in soups, stews, etc. Its cultivation forms quite an important item in market gardens, particularly under glass.

The manner of cultivating it thus is by sowing it between the rows of a growing crop of Lettuce in our cold frames, in April. As it is slow to germinate, it only appears at the time the Lettuce is cut off in May. It is then cleared from weeds, hoed, and forms a growth fit to cut a month before that sown in the open ground. After the first cutting has been made, in June, it is generally so low in price as not to be worth marketing, so it is allowed to grow through the summer until the first week in September, when it is cut off close to the ground and *thrown away*, as it is rarely wanted at this season. It is again hoed, and as at this time it makes a short healthy growth, suitable for keeping well in winter, it is stowed away in narrow shallow trenches, exactly in the same manner in which we preserve Celery.

This way of growing Parsley, I believe, is nearly confined to New York; but as the consumption of such an article is necessarily limited, this market has been oversupplied of late years. Formerly it has frequently paid twice the value of the sash that covered it, in one season—$6 for a 3×6 sash. No doubt, in many places this system of growing would be as profitable as it used to be with us. When not grown under glass, it should be sown thickly in rows a foot apart, in early spring. The varieties cultivated are the "Dwarf Curled," for framing and general crop, and the Moss or Fimbriated, for garnishing.

PARSNIP.—(*Pastinaca sativa.*)

Of late years, our market garden grounds have become too valuable to be used in growing this vegetable, the competition from well cultivated farm lands having brought it down below our paying level. Its cultivation is, in all respects, similar to the Carrot. The soil most suitable is a deep sandy loam, moderately enriched. It is sown rather thickly in our gardens in early spring, at a distance of 12 or 14 inches apart in the rows; on farm lands, at 18 or 20 inches, or wide enough for rows to be worked between by the cultivator. Like all vegetables of this nature, it must be thinned out to a distance of 3 or 4 inches apart between the plants; and our oft repeated caution about weeds must be here again enjoined. It is used almost exclusively in winter, but in our Northern States, what is wanted for winter use, must be dug up in fall, and packed away in the manner described under the head of "Preserving Vegetables in Winter." What are wanted for sale or use in spring, are best kept in the bed where they grow; being entirely hardy in our coldest districts. About one-half is usually dug up and pitted in fall, for sale in winter, and the other half left over for spring. But it sometimes happens that the winter supply is exhausted before the frost is out of the ground in spring, sufficiently to permit of their being dug, and when procurable at such times, they command almost fabulou prices.

On one such occasion my salesman reported that there was not a root of this vegetable to be found in market, and suggested an attempt to dig them at any cost. On an examination of a well sheltered plot, we found it practicable, with crowbars, picks, and wedges, to extract them slowly from the frozen soil, and with our ordinary force, a few barrels were dug that day that were quickly sold at $10 per barrel. I at once secured a supply of extra laborers, and by our efforts the next day, we sent in 40 barrels that sold for $6 per barrel; three or four days more exhausted our supply, but the plot, of little more than half an acre, brought nearly $800, which would not have sold for more than $200, had not the unusual scarcity in market been taken advantage of. The average market price is about $1.50 per barrel, and at that rate, as a farm crop, it is, in my opinion, by 50 per cent. a better paying crop than Onions. It will average easily 200 barrels per acre, and in our rich garden soil about 300. The expense of raising I should judge to be not more than $100 per acre on farm land; in gardens about $200. The increased cost in the garden being mainly in the greater value of the land, for it will be remembered that the annual rent of leased gardens in the vicinity of New York, and other large cities, is about $75 per acre.

A number of varieties of Parsnips are enumerated in seed lists, but the distinctions, as far as I have seen. are hardly worth a difference in name, and I am inclined to think that the soil often determines peculiarities of variety. Certain it is, that by sowing the "Hollow Crowned" on heavy soil, it will be in a great measure deprived of that distinction, while the same seed sown on light sandy soil, will have this peculiarity well marked.

PEA.—(*Pisum sativum.*)

The Pea is grown largely for market purposes in nearly every state in the Union, the time at which it is sown and matures being at widely different dates in northern and southern sections. In any district, its highest degree of perfection is attained under a comparatively low temperature, hence it is one of the many vegetables described as best to be sown in "early spring." True, it is sown for a succession crop throughout the summer months, even as late as August, but the first sowings, everywhere, always produce the best results, and it is from the first sowings only that it is ever offered in market. For market purposes it is more a crop of the farm than of the garden, and many hundred acres are cultivated in Southern Jersey and Long Island for the New York market. Warm, light soils, moderately enriched by stable manure or bone dust, are best adapted to its culture, but if the ground has been manured the previous year, no manure is needed. The whole crop is marketed by July, and is usually followed by a second crop of Late Cabbages or Turnips. The two crops together, average a profit of from $150 to $300 per acre, according to earliness, condition of soil, etc. There is an important matter connected with growing Peas, that confines their culture to the vicinity of a town or village; it is the necessity of being able to get a large number of hands to pick, at the time they are marketable. The variation in one day, in the market, is not unusually from $2 to 50 cents per bushel, which shows the vast importance of an early crop. From the soft condition in which it is required to be gathered, it is a vegetable not very manageable to ship, and the packages, which should be of latticed boxes or baskets, snould never exceed the capacity of a bushel, when shipped from distances requiring from two to three days in the transit. But even this expense and care is well repaid by the high rates for which the first lots are sold. When grown as a market crop, Peas are never staked, and are sown in single rows 2 to 3 inches deep, and from 2 to 3 feet apart, according to the variety, or the strength of the soil. When grown in small quantities for private use, they are generally sown in double rows, 6 or 8 inches apart, and staked up by brush, for the taller growing kinds.

The varieties are very numerous, but are in a great state of confusion, the same kind being often sent out under a dozen names. The following varieties are well defined, arranged as our experience gives the order of merit for this locality.

EARLY VARIETIES.

Daniel O'Rourke—Still stands at the head of all other varieties, for the combined qualities of earliness and productiveness. It is the variety mainly grown for market in this district, and in fact, must be in all parts of the country, judging from the immense quantities of it sold by the seedsmen. It should be sown, for a field crop, in rows from 2 to 2½ feet apart, about 1½ bushels of seed being required per acre.

Extra Early.—We find this to be a few days earlier than the preceding, but not quite so large in the pod, and hence not so profitable for market, but desirable as the earliest sort for private use.

Tom Thumb.—A much valued variety for its extreme dwarfness, which does away with the necessity for stakes; it is, besides, very productive, as it is planted in rows 1 foot apart; it grows from 8 to 12 inches high. It is occasionally grown as a market variety on heavy soil, which is best suited for it. Being planted closer in the rows, it requires at least 2 bushels of seed per acre. It is also a very hardy variety, and is generally used for sowing in fall in the Southern States, where it sometimes, however, requires a little protection by brush during winter; thus sown it matures very early in spring.

Bishop's Long Pod.—Said to be a cross between Bishop's Dwarf and the Marrowfat, partaking of the dwarf and early qualities of the former, with the great productiveness of the latter; one of the very best for domestic use.

McClean's Advancer.—A comparatively new variety, becoming fit for use in a week after the earliest sorts; pod and pea large, and of excellent flavor.

LATE VARIETIES.

Champion of England.—This, so far, is, by general consent, acknowledged as the best of the late varieties. It is tall growing, four feet in hight, requiring to be staked up; pod and peas of the largest size.

British Queen.—Plant strong and vigorous, often attaining a hight of six feet. Aside from its large size, it has the merit of continuing long in bearing, and is less affected by mildew in summer than most other varieties.

Blue Imperial.—One of the oldest varieties, but yet

standing with undiminished merit as one of the best late summer varieties; it is one of the latest, very productive, and of excellent quality.

Veitch's Perfection.—One of the dwarfs of the late varieties, abundantly productive; pods and peas of the argest size; a favorite fall sort.

Black and White-eyed Marrowfat.—These are both productive and hardy varieties, extensively grown as field peas; used dry. They are also, from their great productiveness, grown largely in private gardens, but they are not so fine flavored as most other varieties.

Tall and Dwarf Sugar.—These are the varieties known as "edible podded," and are excellent to use in the green state, in the same way as String Beans, retaining almost the identical flavor of the Pea. When not used with the pods, they are equally excellent as shelled Peas, and as the name implies, particularly sweet.

PEPPER.—(*Capsicum annuum.*)

A tropical plant, that requires to be started in hot-beds or forcing pits, in the Northern States. The most common method is, to sow in hot-beds in March, and treat in all respects as directed for the cultivation of the Egg Plant. Light sandy soils are rather best suited for its growth, but it will grow tolerably well on any soil. When cultivated for market, they are planted in rows 2 feet apart, and 15 inches between the plants. The crop is moderately profitable, but it is not grown in large quantities, the main consumption of it being by the pickle factories.

The popular varieties are;

Bull-nose, or Bell.—An early variety of mild flavor, rind thick and fleshy; it is a favorite variety both for pickling and for use in the crude state.

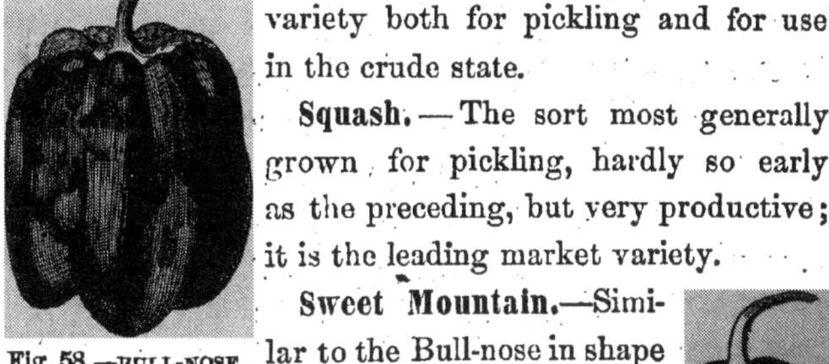

Fig. 58.—BULL-NOSE, OR BELL PEPPER.

Squash.—The sort most generally grown for pickling, hardly so early as the preceding, but very productive; it is the leading market variety.

Sweet Mountain.—Similar to the Bull-nose in shape and color, but larger, and milder in flavor; used to make stuffed pickles.

Cayenne.—The variety of commerce; pods small, cone-shaped; coral-red when ripe; it is quite a late variety, but the pods are as frequently used for pickling green as when ripe. Extremely acrid.

Fig. 59.—CAYENNE PEPPER.

POTATO.—(*Solanum tuberosum.*)

The soil acknowledged to be best suited for the Potato, is sandy loam; in all heavy soils it is more subject to disease, and the flavor is also much inferior. This, however, is true of nearly all vegetables, heavy soils inducing a watery insipidity of flavor. Like all robust growing vegetables, Potatoes can be grown with varying success on soils of all kinds and in all conditions of fertility; but it is every way most profitable to use an abundance of manure, when it is at all attainable. In breaking up good pasture land, the decaying sod answers sufficiently well for the first year in lieu of manure. Manure is applied either in the rows or hills, or broadcast over the surface, and plowed in; the latter plan in all cases being preferable, when manure can be obtained in sufficient quantities.

Potatoes, when grown for market, are always a farm crop, the receipts per acre being much too low for the regular market garden; the large quantities that are planted usually prevent the use of manure in any other way except in the rows. When thus applied, furrows are plowed out in spring, after the ground has become dry and warm, usually 3 feet apart, and from 3 to 4 inches deep. The manure is spread in the furrow, the "sets" or "seed" planted thereon, from 8 to 10 inches apart, and the furrow again covered in by the plow. As soon as the shoots are seen above ground, the ridge should be at once hoed, and the cultivator run between the rows; as they advance in growth, the soil should be laid up on each side against the row, so as to form a slight ridge.

Considerable discussion has at various times occurred concerning the relative merits of planting cut or whole tubers, but is yet undecided, each system having its advocates; a fact which goes far to prove that it is of little consequence which method is followed. The best rule, in our experience is, to plant the whole small tubers if fully matured, and the larger ones cut, but in either case leaving enough bulk to give sufficient sustenance to the plant.

The Long Island farmers, in the neighborhood of New York, have their crops of Early Potatoes sold off early enough in July to get the ground leveled and Late Cabbages planted on the ridge on which the Potatoes have been growing, sufficient manure being left in the ground to carry through the crop of Cabbage. The two crops together give an average profit of $150 per acre. Potatoes may be preserved during winter by the section pit system recommended for the general preservation of vegetables, or in a frost-proof cellar.

The varieties of the Potato are very numerous, many of them having only local reputations, so that it is somewhat difficult to name the best for such an extensive territory as ours; those below named seem to have the most qualifications to commend them for general culture.

Early Goodrich.—A new variety raised in 1860 by the Rev. C. S. Goodrich, of Utica, N. Y., who, from many

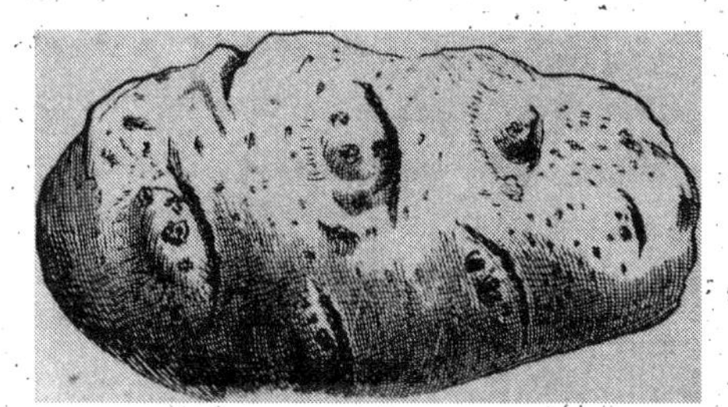

Fig. 60 —EARLY GOODRICH.

thousands of seedlings, selected this as the most meritorious. It has been thoroughly tested since then, and all bear unvarying testimony in describing it as one of the earliest, most productive, and equal in quality to any variety cultivated. The cut represents half the average size.

It also has the excellent property of producing very few small tubers. It should be planted, when in rows, 2½ feet apart.

Early Dykeman.—This has been the leading variety grown in this vicinity for an early crop for the past six or eight years, but there is hardly a doubt but that it will be superseded by the Goodrich just as soon as that variety gets plenty enough for general cultivation. In size, the tuber is above medium, yellowish-white, eyes rather sunk, purplish.

Ash-leaved Kidney.—An old English variety of unsurpassed earliness, dry, and of excellent flavor. Tubers kidney-shaped, rather small; skin yellowish white. This is an excellent variety for early crop for private use, but is not productive enough for market; may be planted from 1½ to 2 feet in the rows.

Buckeye.—Second to the Dykeman as an early market sort, and surpassing it in quality. The tuber is roundish with pink eyes, and above average size. A strong grower and very productive; plant 3 feet apart in the rows.

Jackson White.—A comparatively new variety, of great excellence; tubers large; color yellowish-white; skin often rough. An excellent flavored variety, and though not early, is now extensively grown in this vicinity for general crop; plant 2½ to 3 feet in drills.

Lapstone Kidney.—An English variety of great size and exceeding productiveness; it is quite late, however, and its chief merit is in its quality of keeping late in spring; it is quite a favorite in private gardens here. Tubers large, oblong, flat; color yellowish-white; plant 2½ feet apart in drills.

Mercer.—A variety perhaps more generally known and cultivated than any other sort; it is very productive, medium size, and of excellent flavor, and keeps well until spring. One of the leading market varieties.

Harrison.—One of the Goodrich seedlings. Said to be the most productive of all Potatoes; color yellowish-

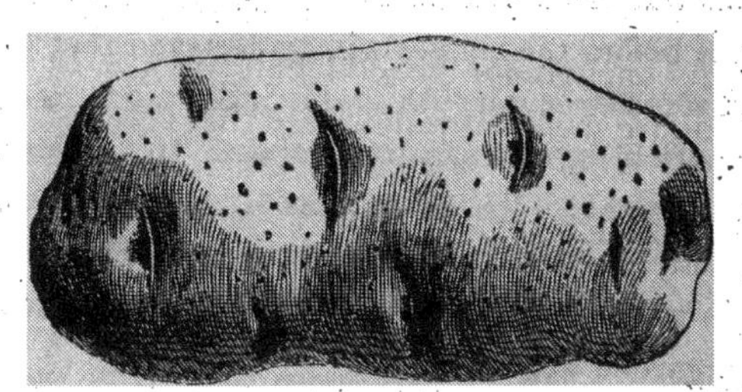

Fig. 61.—HARRISON POTATO.

white; oblong; full; flavor excellent. At this date, the most valued and highest priced on the list. The above engraving represents half the average size.

Peach Blow.—Another sort extensively grown for market, and a favorite shipping variety. Tubers rather large, round; eyes rather sunk; skin pinkish-red. Very productive, and in this section has for some years withstood disease better than any other sort; should be planted 2½ or 3 feet in drills.

White Peach Blow.—A sub-variety of the above, claimed by some to be superior.

PUMPKIN.—(*Cucurbita Pepo.*)

The Pumpkin is yet offered in large quantities for sale in our markets, but it ought to be banished from them as it has for some time been from our gardens. But the good lieges of our cities are suspicious of all innovations in what is offered them to eat, and it will be many years yet before the masses will understand that the modest, and sometimes uncouth looking, Squash is immeasureably superior, for all culinary purposes, to the mammoth, rotund Pumpkin. The Pumpkin is an excellent agricultural plant, of great value for cattle, but I have no reason to allude to it here, except to denounce its cultivation or use as a garden vegetable.

RADISH.—(*Raphanus sativus.*)

Radishes are consumed in immense quantities, and are one of the vegetables which we deem of no little importance as a market crop. To have them early, a light rich soil is the best; heavy or clayey soils not only delay their maturing, but produce crops much inferior, both in appearance and flavor. They are grown by us in various methods; the most common is, after sowing a crop of Beets in rows, to sow the Radish crop regularly over the bed broadcast. The Radishes come up quickly, and are gathered and sold, usually in six weeks from the time of sowing. The Beets at this time have only become large enough to be thinned, and will not be ready for at least a month later, so that the Radish crop is taken off the same ground with little or no injury to the Beet crop. Another method is, to sow them between the rows of Early Cabbages or Cauliflowers, where they also are gathered off so soon as not to interfere with these crops.

These are the methods practised in our gardens here, where land is so valuable that we must make it always carry double, and often treble, crops in a season. Radishes are also grown in some places very extensively, on land devoted exclusively to that purpose, in spring. Their culture thus is exceedingly simple. The ground being plowed and harrowed well, the seed is sown, and the harrow again run over, which places the seed at the proper depth. But though the field cultivation of this vegetable is simple, the labor of gathering, tying up, and washing, preparatory for market, is great, which detracts largely from the profits. Perhaps the average receipts are $300 per acre, but the expense, before this is realized, is probably one-half that amount. It must be remembered, that, in many cases, it is an auxiliary crop, interfering but little with our main spring crops. It is one of the vegetables convenient to ship, and the early samples from Norfolk, Va., average $10 per barrel, of 200 bunches; or about $1000 per acre, which should be a great inducement to southern cultivators, as there is but little danger of glutting the markets with fresh vegetables shipped from a southern to a northernport. Only a few varieties are cultivated, although the seed lists give dozens.

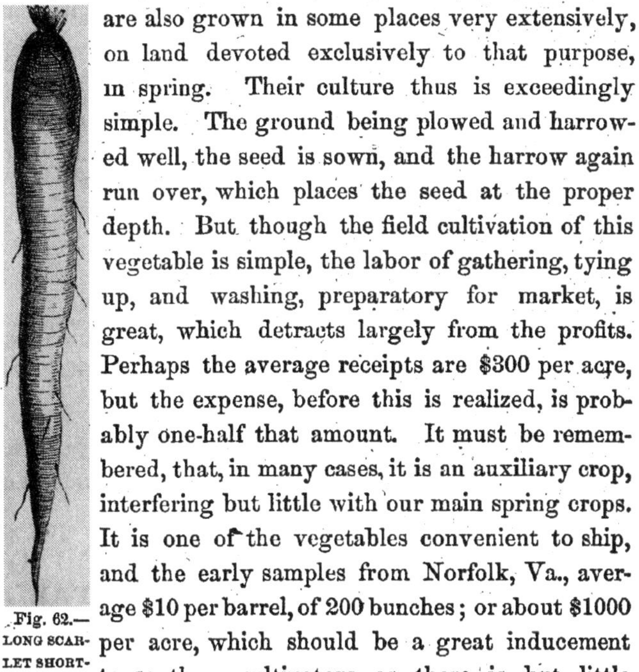

Fig. 62.—LONG SCARLET SHORT-TOP.

Long Scarlet Short Top.—This is the variety grown

in rather the largest quantity, as from its shape, (when tied up in flat bunches), it is best suited to ship. In rich light soils, its average length is about 9 inches.

Scarlet Turnip.—Rather more delicate in flavor than the above, and for this reason in more demand for home consumption. By allowing it sufficient time to grow, it attains 3 inches in diameter, but it is always gathered at half this size. This, and the preceding, are the two varieties that are grown as early market sorts.

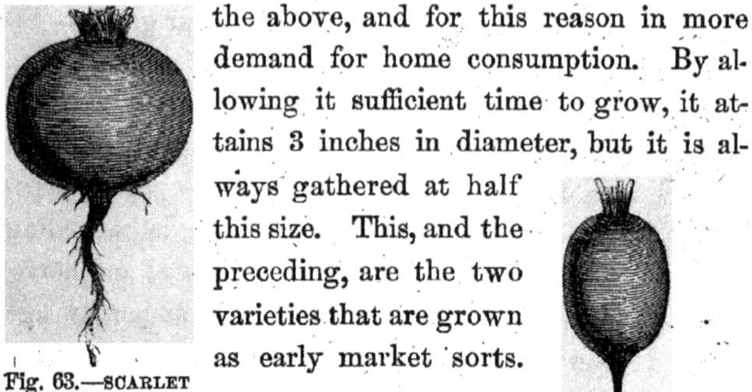

Fig. 63.—SCARLET TURNIP RADISH.

Fig. 64.—OLIVE SHAPED RADISH.

White Turnip and Long White—Are merely sub-varieties of the above, differing in no way except in color; they are generally grown with the above in private gardens for the sake of variety, but seem not to be esteemed in market.

Scarlet Olive-shaped.—An excellent variety; color crimson, rather than scarlet; small top; early, tender, and in every way desirable.

LATE VARIETIES, GROWN FOR FALL AND WINTER USE.

Yellow and Gray Turnip-rooted.—Varieties well adapted for summer use, as they stand the heat better than the early sorts; they are mild in flavor, but are but little grown, as few relish Radishes at that season.

Rose-colored Chinese.—A valuable variety; color pink or rose; skin smooth; of sharp but agreeable flavor. Keeps as well as any.

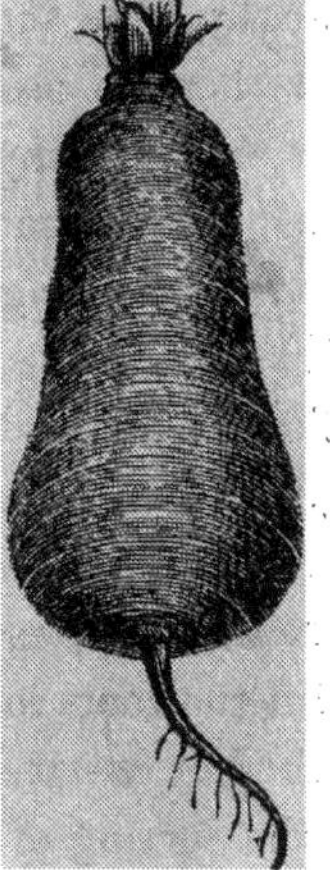

Fig. 65.—ROSE-COLORED CHINESE RADISH.

Black Spanish.—A very hardy variety often preserved, keeps as well as any other root in sand until mid-winter, in cellar or root-house; large size; color grayish-black; rather harsh in flavor.

RHUBARB.—(*Rheum Rhaponticum.*)

Rhubarb is now cultivated largely for market purposes in the vicinity of all large cities, and few private gardens are without it. Its culture is of the simplest kind. It is increased by division of the root, every portion of which that has an eye or bud will form a plant. It is essential, to grow it early and in perfection, that the soil be broken up at least 18 inches deep by the plow and subsoiler, and manured heavily; 100 tons per acre, if attainable, will be none too much. It may be planted in early spring, or in fall, 3 feet apart each way; if planted in spring, on ground well prepared, a full crop can be gathered the succeeding season. It is a vegetable requiring but little labor; once planted, it will remain in bearing condition for three or four years, only requiring a top-dressing of manure, dug in, in spring or fall. It is one of those crops of which it is difficult to state the value per acre; as in the varying conditions of earliness, it is sold from $200 to $1000 per acre. It may be safe to say, however, that it will average, in this vicinity, a profit of $350 per acre annually, from the date of planting; in many places, where there is less competition, no doubt double or treble that amount may be realized. Rhubarb is a most simple and convenient plant for forcing, which may be done either by hooeing the crowns, or inverting barrels over them in early spring, say two or three weeks before the weather opens, and covering the whole bed up to the depth of 3 feet, with eaves or other heating material. Rhubarb requires no light in forcing, the stalks being much more tender when light is excluded. Another method is, to lift the roots in a mass in the fall, and place them in a back shed or furnace room; such situations as are adapted for the forcing of Mushrooms, will force Rhubarb; the requirements being moisture, and a temperature of from 45° to 60°.

Forced Rhubarb always sells at highly remunerative prices, and when there is a surplus of roots, and conveniences for forcing, it will pay handsomely. The varieties mainly cultivated are:

Linnæus.—This is the variety mainly grown by the market gardeners for an early crop; it is also very productive, of excellent spicy flavor, being the least acid of all the varieties. It is small, however, and is superseded by the larger but later sorts as soon as they appear.

Victoria.—This sort is also largely grown for markets, but mainly on lands that are not early; its great size and consequent weight of crop, compensating for its lateness. It is therefore recommended for heavy late lands, and the preceding variety for light and early soils.

Cahoon.—This variety is particularly abundant of juice

and when "wine" is made from Rhubarb, is perhaps the best fitted for the purpose. Rhubarb is not likely to make much headway as a "wine plant," the disinterested (?) efforts of its advocates, notwithstanding.

Early Prolific.—A variety that is extensively and almost exclusively grown by the London market gardeners, but has not yet been much tried with us; if what we have heard of it is correct, it is far in advance of all others as a market sort.

SAGE.—(*Salvia officinalis.*)

See Thyme, and other Sweet Herbs.

SALSIFY, OR OYSTER PLANT.

(*Trayopogon porrifolius.*)

This vegetable is coming rapidly into general use; patches of an acre of extent are seen in some of the more extensive gardens here, that a few years ago grew only a few rows. It should be sown in early spring. Its culture is in all respects similar to that of Carrots, and although its consumption is as yet limited, for what is grown of it, the prices are high and very remunerative. It is moreover a safe root to cultivate, for, being entirely hardy, there is no risk whatever of its being injured by frost, should it so happen that the digging up is neglected in the hurry of fall work. It is generally better, however, to dig it up and put it away as we do Parsnips or Carrots, so that it can be got at any time during winter. It usually sells at higher rates in spring, than in fall or winter; but even with the advance in price is less profitable to the grower when sold in spring, as that being a busy season, the labor expended in digging it up and getting it ready, is then of much more value.

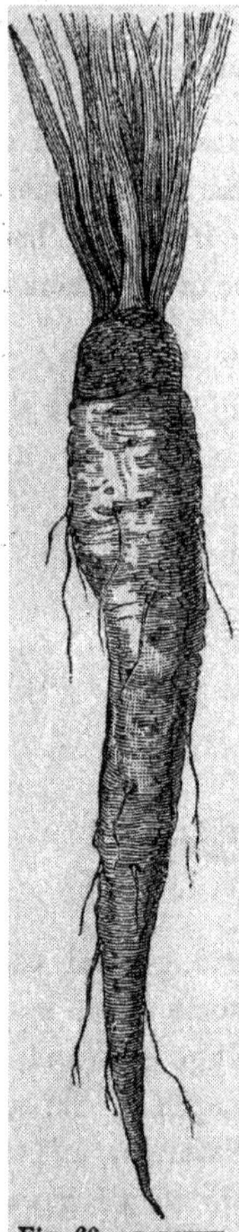

Fig. 66.—SALSIFY.

As this vegetable will be unknown to many, I will state that it is used in various ways, but generally boiled, or stewed, like Parsnips or Carrots. It is also used to make a soup, which has a decided flavor of the Oyster. It is also eaten as a salad, sliced and dressed with vinegar, salt, and pepper. There are no varieties.

SCORZONERA—BLACK SALSIFY.

(*Scorzonera Hispanica.*)

Very similar to the above in general character, and of the same culture and use. It is not, however, so generally esteemed as the Oyster Plant, and is not grown except for private use, and even for that purpose sparingly.

SEA KALE.—(*Crambe maritima.*)

This vegetable is much esteemed and grown largely for market purposes, both in England and France, and no private garden of any extent in either country, is considered complete without it. Here, however, we seem to make but little headway in its cultivation. I have never yet seen it offered for sale in our markets, and its culture is practised in but few private gardens.

There is an impression that it is difficult of culture in our climate; this is not so, by any means; it is equally as easy to grow it here as it is in England, only that, like all vegetables, requiring artificial heat for its perfection, its cultivation is attended with more expense than that of vegetables that we plant in the open ground, without other care than to keep them clear of weeds.

It is increased either by roots or by seed; when roots can be obtained to start with, they are quicker. The manner of operating with them is as follows: in fall, a few old plants of Sea Kale are dug up, and the roots cut in pieces of from 2 to 3 inches in length; these are placed in boxes of sand in a dry cellar, until February or March; they are then strewn on the surface of a hot-bed, where, in a week or two, they will emit roots and tops; they are then potted, hardened off for a few weeks, and as soon as the weather is settled, planted out in rows 3 feet apart, and 2 feet between the plants. If the ground is in the condition that it should be, Sea Kale, thus treated, will produce crowns strong enough to give a crop the next season after planting. When grown from seed, the seed should be sown in hills, at the above distances apart, in early spring, each hill being thinned out to three or four plants. In our colder latitudes, the crowns should be covered by 4 or 6 inches of manure or leaves, as a protection from frost.

Sea Kale is only fit for use in the blanched state, consequently, on the approach of spring, the "crowns" should be covered with some light material, such as sand or leaf mold, to the depth of 12 or 15 inches, so that the young shoot, being thus excluded from the light, will become blanched in growing through this covering, or, sometimes cans, made for the purpose, or inverted flower pots, are used, the only object being to exclude light. In England it is forced extensively, by covering over the whole beds with leaves, manure, or some heating material. The young shoots, when cooked, have a flavor something between Asparagus and Cauliflower, but is much preferred to either. The engraving shows a young shoot, when ready for the table. The fully developed leaves are large and robust.

Fig. 67.—SEA KALE.

SHALLOTS.—(*Allium ascalonicum.*)

This vegetable, which is used in the green state in early spring as a substitute for Onions, is planted by dividing the bulb in September, and planting in rows 1 foot apart and 6 inches between the plants; it makes a slight growth and forms its roots in the fall. On the opening of spring, it developes rapidly, and the single bulb, planted in September, will have increased, by May, a dozen fold. From its hardy nature, coming in at least three weeks earlier than the Onion, large quantities are sold, at rates corresponding with those of Onions raised from sets. It, with us, has ever been a profitable vegetable to raise, and I have arely found the profits on an acre to have been less than $300. It is generally cleared off by the last week in May, giving sufficient time to follow with second crops of Early Cabbage, Beets, Turnips, etc.

SORREL.—(*Rumex acetosa.*)

A well-known perennial plant, cultivated to some extent with us. It is used in soups and sauces, mostly by the Germans and French. In the French markets, it is nearly as abundant as Spinach is in ours, and is highly recommended as a wholesome vegetable. Its cultivation i very simple. Seeds, sown thinly in rows in early spring will give a heavy crop of leaves in June and July; when the flower-stalk of the Sorrel starts to grow, it should be cut out, which will add greatly to the development of the leaves. The crop may be left two seasons, but is more tender when annually raised from seed.

SPINACH.—(*Spinacia oleracea.*)

This is a very important crop in our market gardens, hundreds of acres of it being cultivated in the neighborhood of New York. It is one of the most manageable of all vegetables, requiring but little culture, and may be had fit for use the entire season.

In our market gardens, it is sown in early spring as an auxiliary crop, between the rows of Early Cabbage; it comes to perfection usually in five or six weeks after sowing. At this season, it sells at a low price, usually from 50 cents to $1 per barrel; but it requires but little labor, and generally pays about $75 per acre of profit. The main and important crop is sown in drills, 1 foot apart, in this section from 1st to 15th September, or late enough in fall to get about half grown before cold weather sets in. It is sometimes covered up, in exposed places, with straw or salt hay during winter, which prevents it being cut with the frost; but in sheltered fields, here, there is no necessity for covering.

Thus sown, in the fall, it is begun to be cut or thinned out for market, about 1st of April, and is usually cleared off by 1st of May, giving the ground for a second crop of Cabbage, etc. I could never account for the fact that some vegetables always continue to be more profitable to raise than others that require the same expenditure of labor; here we have a marked case in point. Spinach, which certainly requires no more labor in raising than a crop of Potatoes, continues to give a profit of at least three times as much per acre, on fields divided only by a post and rail fence. The men that grow the Spinach are never foolish enough to encumber their ground with Potatoes;

but their immediate neighbors have done so for the last dozen years, and have never discovered that $50 expended more per acre in manure, would annually put $200 more per acre in their pockets, by growing Spinach, instead of Potatoes.

The varieties are very few.

Round.—This is the variety generally cultivated for winter use, being remarkably hardy, and standing our severest winters with but little injury. It is the main market sort.

Prickly.—Although this variety is usually sown in spring and summer, it also stands well in winter, but gives less bulk per acre than the first named.

New Zealand Spinach.—(*Tetragonia expansa.*)—A plant of the same character and uses, but of a different genus, and used only in private gardens. It is a remarkable plant, of low branching habit, growing with wonderful luxuriance during hot weather. Single plants often measuring 6 feet in diameter. The leaves are used exactly as common Spinach; it is best grown by sowing the seeds in April and May, and transplanting to 3 feet apart.

SQUASH.—(*Cucurbita Pepo, and C. maxima.*)

A class of vegetables embracing more marked distinctions in sorts, fitted for more varied uses, and to be found, during the extremes of the season, in a better state of perfection, than, perhaps, any other product of our gardens. Being of tropical origin, their growth is all consummated during summer; yet the fruit of the "winter varieties" may be kept, with a little care, until May. They are all of luxuriant and vigorous growth, and although they will grow readily on almost any soil, yet there is hardly anything cultivated that will so well repay generous treatment. Like all plants of this class, it is useless to sow until the weather has become settled and warm; next to Lima Beans, Squashes should be the last vegetable planted. Light soils are best suited for their growth, and it is most economical of manure to prepare hills for the seeds, in the ordinary manner, by incorporating two or three shovelfuls of well rotted manure with the soil, for each hill. For the Bush varieties, from 3 to 4 feet each way, and for the running sorts from 6 to 8 feet. Eight or ten seeds should be sown in each hill, thinning out after they have attained their rough leaves, leaving three or four of the strongest plants.

They are extensively grown for market, but are not sufficiently profitable for our highly cultivated gardens, and are therefore grown rather as a farm-garden crop. They give a varying profit, in our vicinity, of from $100 to $10 per acre. The early varieties are grown quite extensively in the vicinity of Norfolk, Charleston, and Savannah, and shipped North, from two to four weeks earlier than they can be had here, and like all such commodities, bring three or four times the price of those grown in the vicinity, in quantities that glut the market.

The varieties are very numerous, and from the facility with which they will cross, it is very difficult to retain the different kinds pure.

SUMMER VARIETIES.

Yellow and White Bush Scalloped.—These are the two varieties that are esteemed the most early, and are such as are almost exclusively grown for market, for the first crop; from the hard texture of the rind, they are well fitted for shipping, and are the sorts grown exclusively at the South for that purpose. The characters of these varieties are very decided, never presenting any variation. Plant 3 to 4 feet apart in hills.

Fig. 68.—WHITE-BUSH SCALLOPED SQUASH.

Summer Crook-neck.—A much esteemed variety in private gardens, somewhat similar in growth to the Bush; rather more dwarf. The fruit is orange-yellow, covered with warty excrescences; usually from 7 to 9 inches long; considered the best flavored of the summer varieties.

Boston Marrow.—This variety may be termed second early, coming in about ten days after the Bush and Crook-neck sorts. The skin, which is of a yellowish shade, is very thin; the flesh thick, dry and fine ground, and of unsurpassed flavor.

FALL, OR WINTER VARIETIES.

Hubbard.—A general favorite, and more largely grown as a late sort than any other; it is of large size, often weighing from 9 to 10 lbs. Color blueish-green, occasionally marked with brownish-orange or yellow; flesh fine grained, dry, and of excellent flavor. It can be had in use from September to May, eight months of the year. It should not be planted closer than 8 feet apart.

Yokohama.—A very distinct variety sent from Japan, by Mr. Thomas Hogg, in 1860, and since very generally disseminated. The fruit is roundish, deeply ribbed; color orange-salmon; thickly warted; flesh thick, very dry, sweet, and excellent; maturing earlier than the Hubbard, but not so desirable for winter. Plant 8 feet apart.

Winter Crook-neck.—A variety largely grown in some of the Eastern States, where it is said to be kept the entire season. Skin reddish-pink, when matured; flesh close grained, sweet, and fine flavored. It is a luxuriant growing variety, and should be planted in hills 9 feet apart.

Vegetable Marrow.—This variety is a favorite English sort; the fruit is very variable in size, from 9 to 18 inches in length, by from 4 to 6 inches in diameter. The skin is greenish-yellow; flesh white, soft, and of rich flavor; very distinct from all of the preceding. Plant in hills, at 8 feet.

SWEET POTATO.—(*Ipomœa Batatas.*)

We have few vegetables that are so particular about soil as the Sweet Potato, and to succeed well with it, it is essential that the soil be light, rich and warm. It is labor lost to attempt its cultivation on a heavy, cold soil. It is more generally grown in the Southern States than the common Potato, as there the soil and climate are more congenial to it. We have often difficulty, in this district, in saving the tubers sound enough until spring, to start for sprouting to produce young plants. The great essentials to their good preservation, are a dry and rather warm atmosphere; the cellar, suitable to preserve the common Potato, being usually much too cold and damp for this. Where there is no place of the necessary high temperature, it is best to get them in spring direct from some southern market, where they can always be had in good condition; or they can be kept by packing in barrels in dry sand, and keeping them in a warm room. In this district, we begin to start the tubers in hot-beds or forcing pits, about the middle of April, laying them thickly together on a 2-inch layer of sand and leaf-mold composted together, (or sand alone will suit if leaf-mold cannot be had); as soon as the buds or eyes show signs of starting, cover the tubers completely over to the thickness of an inch with the same material. Treat as for other tender plants in the hot-bed or forcing pit, and the sprouts or slips will be ready for planting out by the first of June.

Market gardeners often make the sale of Sweet Potato plants a very profitable operation, immense quantities of them being sold to private growers at the planting season. As the sprouts from the tubers come up very thickly, repeated thinnings are made, which is not only profitable to the grower, but of great advantage to the remaining plants, by giving them the necessary room to grow. One grower, in this vicinity, informed me that last season he sold upwards of $1000 worth of plants from 150 sashes. The profit from the cultivation of the plant in the field is something less than that from Tomatoes, but more than from the common Potato.

In this latitude, the Sweet Potato should never be planted much earlier than the first of June; it is very susceptible of being chilled, and the weather is rarely settled and warm enough to be safe here before June. Prepare the hills as for Tomatoes, 4 feet apart, planting three plants in each, or if in rows or ridges, 4 feet apart, and 1 foot between the plants; in either case requiring from 8,000 to 10,000 plants per acre.

The following are the sorts mostly grown.

Nansemond.—This is the earliest sort; tubers large, from 3 to 4 inches in diameter at the thickest part, tapering to each end, and from 5 to 8 inches long; flesh dry sweet and well flavored.

Red Skinned.—This variety is claimed to be hardier than the preceding, but it is doubtful if this is the case. It is a long, slender variety, mostly grown in private gardens, and is believed to be of a richer flavor than the yellow or white sorts.

Yellow Skinned.—This sort is mainly cultivated in the Southern States, where it attains nearly the weight of the Nansemond; it requires a longer season than that variety, and is not so suitable for the North. It is of excellent flavor, and more free from stringiness than any other sort.

TOMATO.—(*Lycopersicum esculentum.*)

This vegetable is one of the most important of all garden products; hundreds of acres are now planted with it in the vicinity of all large cities, and the facility with which it is managed, places it readily under the control of the least experienced. It is now grown here almost entirely by those who grow Peas, Potatoes, Melons, and other crops of the "farm gardens," as our market gardens proper are too highly enriched and much too limited in extent to render the cultivation of the Tomato profitable. To produce early crops, the seed must be put down in hot-beds or forcing pits, about ten weeks before the plants are safe or fit to put in the open ground. Thus, in this district, we sow in a hot-bed about the first week in March; in April, the plants are fit to be set out, at a distance of 4 or 5 inches apart, in another hot-bed. They are grown there (proper attention being given to the hot-beds as directed under that head) until the middle of May, when they are safe to place in the open ground. They are planted, for early crops, on light sandy soil, at a distance of 3 feet apart, in hills, in which a good shovelfull of rotted manure has been mixed. On heavy soils, which are not suited for an early crop, they should be planted 4 feet apart. Some attach great importance to topping the leading shoot of the Tomato, so that it branches, arguing that by this means we get an earlier and heavier crop; all our experience shows that no benefit whatever is derived from the practice. Like all vegetables grown on so large a scale, and in such varying soil and climate, the Tomato sells in our markets at prices varying widely, from $10 down to 25 cents per bushel. The average price for those raised in the district, being about $1 per bushel. The quantity raised per acre is about 400 bushels. This may seem at first glance to be quite a profitable crop for a farmer; but, every acre necessitates the use of at least

100 sashes, for, on the second transplanting, about fifty plants only can be grown in a sash, and about 5000 plants are required for an acre. On one occasion, having a very suitable soil, I grew about four acres of Tomatoes for three years, which realized me from $1500 to $2000 annually in *receipts;* but I discovered that the operation was a losing one, as, to raise 20,000 plants for my four acres, I had to make use of 400 sashes, in which, in rather less time and with far less labor than it took to grow the Tomato plants, Lettuce could have been grown that would have sold for at least $2 per sash. Thus I lost annually, in preparing for the Tomato plants, half the receipts of the crop, before ever they were even planted. But there are many parts of the country where Lettuce, thus forwarded, could not be sold, while Tomatoes could, which would materially change the aspect of the operation. In the southern sections of the country, convenient to shipping, Tomatoes are largely grown for the northern markets, and there sold at prices highly remunerative to the grower. In many instances, in the Southern States, the cultivation of Tomatoes for market is carelessly done, the seed being merely sown in the open ground and replanted, as we grow Cabbages. No doubt, by starting in January or February with the hot-beds, or even cold frames, and planting out in March or April, they could be had at least two weeks earlier than they are now sent to us.

There are always some one or more varieties, said to be earlier than others, sent out every spring, but it must be confessed that the varieties that we cultivated twenty years ago are not in earliness a day behind those issued as "vastly superior" in 1866. Last spring, to test them thoroughly, I planted twenty-five plants each of the four most popular sorts, under circumstances exactly similar in all respects; there was no difference whatever in earliness, and but little perceptible difference in productiveness.

I believe that our ordinary methods of saving Tomato and all other seeds have, in fact, much to do in preventing us making any advance in procuring early varieties; if we would only take the trouble to always select the first matured fruits, and the best specimens only, for seeds, and so continue, there is no question whatever, but it would amply repay the trouble. But the grower for market grudges to give up his first basket of fruit, that may realize him $5 or $10, for a few ounces of seed, knowing that he can get plenty when his crop is not worth the gathering for market.. But, depend upon it, he makes a mistake, for the seed from his first fruits would, perhaps, pay him a hundred times better, if used for seed, the next year, than any price he might get for it in the market.

In private gardens, where space is often limited, a greater quantity of fruit will be obtained by elevating the branches of the Tomato from the ground with brush, such as is used for sticking Peas, or by tying to laths nailed against a board fence; or, what is neater yet, the hoop training system as practised in France. But for market purposes, on a large scale, it would require to much labor.

The following are a few of the many varieties grown.

Early Smooth Red.—A very old variety, but one which, for general crop for market purposes, I believe is yet unsurpassed. In the trial above referred to, it was tested with the *Cook's Favorite*, *Tilden*, and *Powell's Early*, and with no perceptible advantage in earliness to either, but the Early Smooth presented the greatest

Fig. 69.—EARLY SMOOTH RED TOMATO.

amount of marketable fruit. This variety is of medium size; rich coral-red; roundish; much flattened; very solid.

The Cook's Favorite.—Differing but little from the preceding, except in shape of the fruit, which is rounder and less flattened. It is grown largely of late years in the neighborhood of Philadelphia, and in southern New Jersey, but is not yet a "favorite" in New York.

Tilden.—This variety, issued last season with a fame that had been widely sounded as being far in advance of all others in earliness, productiveness, and every other good quality, will not sustain its recommendation. It proves, with me, no better in any respect than the first named on this list, and if no better, should never have been sent out.

Powell's Early.—Rather a distinct variety with smaller foliage than any of the preceding; fruit smooth round, inclined to cluster; of a deep scarlet crimson color; very solid and with few seeds.

Fejee.—The fruit of this variety is of the largest size; color reddish-pink; very solid, and well flavored; a most abundant bearer. Its lateness, as well as its objectionable color, make it of little value as a market sort.

Large Red.—Fruit large, irregular, but very solid; this variety is preferred by many to the smooth sorts, the flesh being more solid. It is not quite so early, but is preferable when weight of crop is wanted for catsup, or preserving; for this purpose it is grown largely as a late crop.

Large Yellow.—Differing only from the preceding in color.

Red and Yellow Plum.—Beautiful varieties, never exceeding 2 inches in length, by 1 inch in diameter. They are mainly used for pickling and preserving.

Tree Tomato.—This variety is entirely distinct from all the others, in its upright and tree-like habit. It was introduced to this country some six or seven years ago, from France, but has never shown any quality deserving general cultivation, and is now only grown in private gardens more as a half useful curiosity than anything else.

TURNIP.—(*Brassica campestris.*)

The cultivation of the Turnip for an early crop for market purposes, sold bunched in the green state, is in all respects the same as detailed for Early Beets. The profits of the crop are also similar. The Turnip, however, for early crops, is rather more particular about soil than the

Beet, and can best be produced early on light sandy or gravelly soils, highly enriched with manure.

For late crops, sowings may be made, for Ruta Bagas, from May to September, in the different sections of the country; here, the finest roots are obtained by sowing about first week in June. For white and yellow varieties, as they come quicker to maturity, sowing should be delayed four or five weeks later. Here, we sow from the middle of July to the middle of August.

Turnips, whether for early or late crops, should always be sown in drills, about 14 or 18 inches apart. In large quantities, they are sown by the machine, when one pound of seed will be enough for an acre. In the Northern States, it is necessary to take them up on the approach of severe weather, when they are best preserved during winter by being pitted, as recommended for other roots. The late crops of Turnips are by no means so profitable as the early, rarely realizing to the grower more than $75 per acre; but like most other late crops of the garden or farm, they can be grown with less manure, are less perishable if not immediately sold, and are consequently grown by the farmer on his less valuable but more extensive grounds.

The following are the leading varieties grown.

Purple Top Strap-leaved.—The variety mainly grown for early crop, attaining, when well grown, a diameter of 5 to 6 inches, but is first gathered for market use at about half that size. It is a firm, solid variety, free from sponginess, of very handsome appearance; the lower two-thirds of the root is white, while the upper portion has a well defined line of purple.

White Dutch.—An old favorite sort, having nearly the same shape as the Purple Top, but entirely white; it is equally early, and by some thought to be the best of all in flavor; but is less salable in market, its appearance being less attractive.

Yellow Aberdeen.—This is an excellent variety for culinary use, though not so early as the preceding; the bulb is nearly round, of a dull yellow beneath, and purple or green at top. It is a very solid variety, keeping well throughout the winter, and as it attains a greater weight under favorable conditions, is much grown for stock.

RUTA BAGA, OR SWEDES TURNIP.

Improved American.—The leading variety of this division of the family; is grown very largely for winter sale in our northern markets. Under different culture, it assumes a great difference of shape and size, average specimens being 6 inches long, by 4 or 5 inches wide. It is always hard and solid, and is perhaps specifically heavier than any other vegetable root cultivated.

Laing's Purple Top.—A rather late variety, but, in good land, giving enormous crops. It produces a great abundance of leaves, and for this reason requires at least a space of 18 inches between the rows, and 12 inches between the plants. In shape, it is nearly round; smooth skinned, and handsome.

THYME, SAGE, SUMMER SAVORY, AND MARJORAM.

I believe the cultivation of Sweet Herbs, for market purposes, is but little known in this country, except in the vegetable gardens in the vicinity of New York; there it is practised to an extent of perhaps 60 or 70 acres, a fair average product of which would be about $500 per acre. Like the crops of Celery, Spinach, or Horseradish, they are grown only as a second crop, that is, they are planted in July, after an early crop of Peas, Cabbages, Beets, or Onions has been sold off. The varieties are Thyme, Sage, Summer Savory, and Sweet Marjoram, the former two being grown in the ratio of ten acres to one of the others.

The seed is sown in April in rich mellow soil, carefully kept clean from weeds until the plants are fit to set out, which may be done any time that the ground is ready from middle of June until end of July. As the plants are usually small and delicate, it is necessary that the ground be well fined down by harrowing and raking before planting. The distance apart, for all the varieties, is about the same, namely, 12 inches between the rows, and 8 or 10 inches between the plants; the lines are marked out by the "marker." (This is the "marker" used for many other purposes; in lining out the rows for Early Cabbages, for instance, every alternate line is planted, thus leaving them 2 feet apart, their proper distance.) In eight or ten days after the herb crop has been planted, the ground is "hoed" lightly over by a steel rake, which disturbs the surface sufficiently to destroy the crop of weeds that are just beginning to germinate; it is done in one-third of the time that it could be done by a hoe, and answers the purpose quite as well, as deep hoeing at this early stage of planting is perfectly useless. In ten or twelve days more, the same operation is repeated with the steel rake, which usually effectually destroys all weeds the seeds of which are near enough to the surface to germinate. We use the steel rake in lieu of a hoe on all our crops, immediately after planting, for, as before said, deep hoeing on plants of any kind *when newly planted*, is quite unnecessary, and by the steady application of the rake, weeds are easily kept down, and it is great economy of labor *never to allow them to start.* The herb crop usually covers the ground completely by the middle of September. Then, every alternate line is cut out, each plant making about two "bunches." The object in cutting out the lines alternately is, to give room for the remaining lines to grow; in this way nearly double the weight of crop is taken off the ground than if every line had been cut, and it frequently happens, on particularly rich soils, that at a second cutting every alternate line is again taken when the remaining lines, now standing 4 feet apart, will again meet. I had about an acre of Thyme treated by this process, in the fall of 1864, that sold for over $2000,—but this was an exceptional case, the crop being unusually fine, and prices at that time were nearly double the ordinary ones. As before stated, the average yield is about $500 per acre. Herbs are always a safe crop for the market gardener; they are less perishable than anything else grown, for, if there be any interruption to their sale in a green state, they can be dried and boxed up and sold in the dry state, months after, if necessary. The usual price is from $10 to $15 per 1000 bunches, and we always prefer to dry them rather than sell lower than $10 per 1000, experience telling us that the market will usually so regulate itself as to handsomely pay for holding back the sale. The cost of getting the crop raised and marketed will average about $150 per acre, the principal expense being in tying it in bunches. But with many of our in dustrious German gardeners it does not cost half that, as the tying up is usually done by their wives and children in the evenings; a pleasant as well as profitable occupation.

There are but few varieties of the different kinds of Herbs, but of *Thyme* there are several, and it is very important to plant only what is known as the "spreading variety;" an upright sort, sometimes sold in the seed stores, is worthless as a market crop. The variety of *Sage*, known as the Broad-leaved, is the best.

APPENDIX.

PROPAGATION OF PLANTS BY CUTTINGS.

Although this subject is somewhat foreign to a work on vegetable gardening, yet it may be useful to many into whose hands this book may come, to know on what conditions, slips, or cuttings, form roots. The green-houses or forcing pits, shown in another part of the book, are such as can be used with complete success in rooting cuttings of Grape Vines, Roses, Bedding Plants of all kinds, Evergreens, etc., etc.

I do not know that I can present anything original on a subject that has been so often discussed; but, although I have but little that is new to offer, I will endeavor to simplify what too many gardeners, either intentionally or through ignorance, try to surround and befog with mystery.

It is a general belief among many nurserymen that cuttings can best be rooted on benches formed over tanks, but our large experience with every mode of heating induces us to believe that these are not indispensable. I will only say in this connection, that any one who understands the conditions under which cuttings root, can accomplish the work by a hot-bed, or along the front bench of a green-house, with the flue or pipes running underneath, with perfect success; although he could do so more rapidly and with less attention in a Propagating-house, fitted up with all the "modern improvements." While, on the other hand, the gardener that goes to work without a knowledge of these conditions, though provided with the best Propagating-house that ever was planned, will most certainly fail, or, at least, will not have that unvarying success that the man who knows his business ever should have.

Propagation by cuttings is always most successful between the months of October and April, from the fact that during that period we have the necessary low atmospheric temperature, which I will endeavor to show is necessary to complete success.

Our favorite system of propagating is by using cuttings of the "young wood," that is, young shoots that are formed by starting the plants in a green-house temperature, averaging from 40° to 60°. The proper condition of the cutting is easily determined by a little experience. In the case of Roses, the best are "blind shoots," that is, the short shoots that do not show flower-buds; and the time when hey are of the proper degree of hardness is determined by the flower-buds on the plant just beginning to develope. But with bedding plants, generally, we never can get the cuttings too soft, provided that they have not been grown in a high temperature, and a close atmosphere. The tops of the young shoots are always best, although, if an elongated shoot is soft enough, it may be cut into sections of one or two inches in length.

In making cuttings, preparatory to being inserted in the sand of the bench, it is of no importance whatever to cut immediately below a joint, as three out of every four of the gardeners we meet still think necessary.

In making cuttings, our custom is entirely the reverse of that practice, as we cut usually as much below a joint as the cutting is inserted in the sand, — generally something less than an inch. This is done as a matter of economy, both of time and material, as it is much quicker done, and more cuttings can be so obtained than by cutting at a joint; they are also easier planted in the sand: for in putting in cuttings of any kind we never use a "dibber," we merely push the cutting down to the first leaf, when hard enough to bear it; when too soft, lines are marked out in the sand by a thin knife, so that the soft cuttings may be inserted without injury; they are then watered with a fine rose, which compacts the sand sufficiently firm.

I now come to what I have long considered as the only "secret" of successful propagation, namely, the *temperature;* very simple to give a rule for, but still somewhat difficult to keep to that rule without too much variation.

Soft cuttings, or cuttings of the young wood, should have a *bottom heat* of from 65° to 75°, and the *atmosphere* of the house should be always, when practicable, from 10° to 15° *lower*. If this is strictly adhered to, you are just as certain of a crop of healthy rooted cuttings, in from ten to twenty days, as you would be of a crop of Peas or Radishes in May. But once let these conditions be deviated from, for a single hour, by allowing a dash of sun to raise the temperature of the house or frame to 85° or 90°, then the soft unrooted slip will "wilt," its juices being expended, the process of rooting is delayed, and, if the "wilt" has been severe enough, entirely defeated. The same caution is necessary in applying the "bottom heat," for, if the fire is applied indiscriminately, without regard to the weather, it will be found that you will run the temperature of the bench above "the point of safety," (75°), and in proportion as this has been exceeded, so in proportion will be your want of success. It is true that some cuttings will stand a higher temperature than 75° bottom heat, (grape vines, perhaps, 10° more), but with plants in general, it will be better to let 75° be the maximum.

In the propagation of Roses, etc., by cuttings of the old or hard wood, less attention is required, but success is not always so uniform, nor, in my opinion, are the plants so obtained quite so good as those made from cuttings of young wood. We prefer to place old, or hard-wood cuttings, in the north or west side of a house, or, in fact, anywhere where they can be kept the coolest without being actually frozen. Any attempt to apply bottom heat to the degree used for soft cuttings, will almost certainly destroy them. The temperature of the house may range from 40° to 60°.

In propagating grape vines, however, this rule does not apply, as it does to the hard wood of Roses, and other shrubs; with these, the treatment is nearly in all respects similar to that already described in propagating from young wood. The vine delights in a high temperature, and consequently even the eyes or cuttings, in a dormant state, when put in to propagate, require a temperature that would be quickly destructive to the hard or old-wood cutting of a Rose. Grapes, when raised under glass, are always propagated from single eyes, that is, one bud, with about 2 inches of the under part of the shoot attached; these are planted in the sand of the bench, at from 1 to 2 inches apart, (according to the size of the eye or cutting), and pressed down so that the bud is just above the surface of the sand. The eyes may be put in from January until May, but the best season to begin is about 1st March.

I will now say a word in relation to the *sand* or *compost* used for propagating cuttings. I know there is considerable difference of opinion on this subject; almost every propagator having his preferences. My opinion is, that the color or even the texture of the sand or compost has little or nothing to do with the formation of roots; experiments having satisfied me, beyond all doubt, that the sand or compost is only a medium to hold the moisture.

Experiments with *pure water*, *saw-dust*, *charcoal*, *anthracite*, *brick-dust*, and *sands* of all *colors* and *textures*, showed that cuttings placed in each, in the same temperature, rooted almost simultaneously, and equally well. There are rarely ever any deleterious substances in sand, unless it is the saline matter in that taken from the sea shore, which had better never be used when it can be had from anywhere else. Many of my nurseryman friends I know have been victimized to a ridiculous extent in this matter, by freighting sand hundreds of miles to suit the caprice, or *temporarily* hide the failures, of their propagators;

as, for the want of success in two cases out of three, the sand is made the scape-goat.

The most insidious enemy of the young cutting is the spider-web-like substance, which now, by common consent among gardeners, is called *The Fungus of the Cutting Bench.* Whenever this pest is seen, it may be taken for granted that the temperature has been ***too high***, and the atmosphere too close. The remedy is to raise the sashes enough to allow the exit of the heavy atmosphere, which will at the same time *lower* the temperature. I have observed that the "fungus" can never exist to an injurious extent in a temperature below 50°.

Before closing, I will briefly advert to a simple process of rooting cuttings, which is by far the most covenient for amateurs or for professional gardeners, who have no regular propagating-house. It is what is known here as the "Saucer System." It consists simply in filling plates or saucers with sand, the cuttings are then inserted, somewhat closely together—from an inch to two inches apart; the plates are then watered, so that the sand gets into a half-liquid state; they are then placed in the parlor window, or stage of the green-house, ***entirely exposed to the sun, and never shaded.*** All that is further required is, that the sand must be kept in the ***condition of mud*** until the cuttings are rooted, which will be in from ten to twenty days, according to the temperature, or state of the cutting. Great care must be taken that they never get dry, or the whole operation will fail. This is a very safe method of rooting cuttings, and one that during hot weather is preferable to all others.

MONTHLY CALENDAR.

The success of all garden operations depends upon preparatory measures; for this reason, the beginner in the business can be much benefited by being reminded, as he goes along, of the work necessary to be done to ensure successful results in the future. To do this, I must to some extent repeat directions given in the body of the work, but as they will be presented here in a condensed form, they will not tax the time of the reader. As in all other references made to dates, the latitude of New York is taken as a basis, that being not only the point from which our experience has mostly been derived, but also one that will best suit the majority of readers throughout the country. Those whose location is more southerly or northerly must use their judgment in adapting the directions to suit their locality.

JANUARY.—Vegetation in our Northern States is com pletely dormant during this month, so that, as far as operations in the soil are concerned, it might be a season of leisure; but the business of gardening being one that so largely requires preparation, there is always plenty to do.

The ground being usually frozen, and giving us good hauling, it is always the month in which our energies are given to getting manure, muck, lime, etc., into convenient places for spring work. Care should be taken to get manure in heaps large enough to generate sufficient heat to prevent its being frozen, so that it can be turned and broken up thoroughly before it is spread upon the ground. This work is often very slovenly performed, and the value of manure much reduced by inattention to turning and breaking it up during winter. Sometimes it is injured by being thinly scattered, so that it freezes solid; and again, if thrown into large heaps, and left unturned, it burns by violent heating, getting in the condition which gardeners call "fire fanged." It is always an indication that the manure heap needs turning when it is seen to emit vapor, no matter how often it has been turned previously, for it should always be borne in mind that it quickly loses by heating, while it always gains by a thorough breaking up in turning.

January is usually the month in which we have our heaviest snow storms, which often entail on us an immense amount of necessary, though unprofitable labor, not only in clearing roads, but also in clearing off the snow from our cold frames and forcing pits, for even at this season of dormant vegetation, light is indispensable to the well-being of our vegetable plants; *unless they are in a frozen condition*, that is, if we have had a continuation of zero weather, all our plants of Cabbage, Lettuce, Cauliflower, etc., are frozen in the cold frames; if *in this state*, the glass is covered up by snow, it is unnecessary to remove it even for two or three weeks, but if the weather has been mild so that the plants under the sashes have not been frozen when covered by snow, then the snow must be cleared from the glass as soon as practicable. In the green-houses, hot-beds, or forcing pits, where artificial heat is used, the removal of the snow from the glass is of the utmost consequence.

If not done in December, the final covering up of Celery trenches, root pits, and all things requiring protection from frost, should be attended to in the first week of this month.

Should the ground be open enough to allow of digging, (which occasionally occurs here even in January), let all roots, remaining in the ground, be dug up and pitted, as another chance is not likely to occur before spring. Cold frames and forcing pits, particularly the former, should be aired whenever the weather will permit, for it is necessary, to carry them safely through until spring, that they may be properly hardened. (See article on Cold Frames.)

FEBRUARY.—The gardening operations differ but little from those of January, except that in the latter part of the month, as the days lengthen and the sun gets brighter, more air may be given to framing and forcing pits. Hot manure should now be got forward to be prepared for hot-beds, and if desired, some may be formed this month. (See article on Hot-beds.) Have all tools purchased or repaired, so that no time may be lost in the more valuable days of next month. It is important to have always spare tools of the leading kinds, so that men may not be thrown idle, at a hurried season, by the breaking of a fork, spade, or hoe. In harness and implements, connected with the teams, it is of great importance to have spare parts to replace those liable to be broken; otherwise, frequently half a day is lost, by the breaking of a whiffle-tree, or plow share, causing more loss by delay, than three or four times the cost of the article.

MARCH—is one of the busiest months in the year with us. Hot-beds are made, and planted or sown, and Lettuce crops may be planted in cold frames and forcing pits, (see directions under these heads). In the latter part of the month we often begin, on dry soils, the sowing or planting in the open ground of such hardy vegetables as Horseradish, Cabbage, Lettuce, Onions, Radishes, Turnips, etc., etc. Although we gain but little in earliness by starting before April, yet it forwards our operations, so that it equalizes labor more than when starting late in spring. Enthusiastic beginners must avoid the too common error of beginning out-door operations too soon, when the soil is not sufficiently dry; for, if the soil is dug or plowed while wet, it is highly injurious, not only destroying the present crop, but injuring the land for years after.

New plantations of Asparagus, Rhubarb, Sea Kale, and Artichokes may be made, and old beds top-dressed, by digging in short manure close around the plants; we consider it more economical of manure to do this in spring, than in fall. (See Asparagus.)

Such roots as Cabbage, Carrot, Celery, Leek, Lettuce, Onion, Parsnip, etc., planted to produce seed, may be set out the latter part of this month, on soils that are warm and dry, drawing earth up around the crowns so as to protect them from sharp frosts; in hoeing, in April, this soil is removed.

Where extra laborers are wanted for the garden, I have always considered it economy to secure them in the early part of March, even a week or two before they are really needed, for if the hiring of them is delayed until the rush of work is upon us, we often have to pay higher rates for inferior hands, and have less time to initiate them in their duties. To such as require large numbers of hands, and look to such ports as New York for emigrants, let me caution my friends from the rural districts not to believe too implicitly in the promises of these prospective American citizens. Much vexatious experience has taught me that one out of every three men is either worthless, or will run away, so that for many years back, if I wanted four hands, I made one job of it and hired six, well knowing, that before a week had passed, my force would be reduced to the required number.

APRIL—brings nearly all the operations of the garden under way, the planting and sowing of all the hardy varieties of vegetables is completed this month. (See table, in article on Seed Sowing). Look well to the hot-beds, cold frames, or forcing pits; they will require abundance of air, and, (where artificial heat is used), plenty of water; we have now bright sunshine, promoting rapid vegetation under glass, and to have heavy crops, they must not be stinted in water. Hot-beds are particularly critical in this month; an hour or two of neglect, in giving air, may quickly scorch the tender plants that you have been nursing with so much care for a month previous; and a balmy April day may terminate in a stinging frost at night, making short work of your hot-beds if they are not well covered up by straw mats.

Plantations of Asparagus, Rhubarb, etc., if not made last month, should now be done, as those set out later than April, will not make such a vigorous growth. Succession crops of Lettuce, Beets, Cabbage, Onions, Peas, Potatoes, Radishes, Spinach, Turnips, etc., may be planted or sown during the latter part of the month, to succeed those planted in March and early part of April.

The early sown crops should be hoed, and the ground stirred close to the young plants, so as to destroy the germ of the weeds now appearing.

May.—Although the bulk of the hardy vegetables is now planted, yet the tender varieties are still to come; they require more care as they are more susceptible of injury, by too early or injudicious planting, than the others. In the early part of the month, the succession crops, named in April, may be yet planted so as to produce good crops, and the tender varieties, such as Bush Beans, Corn, Melon, Okra, Pepper, Squash, Tomato, may be sown or planted after the middle of the month; but Egg Plants, Sweet Potatoes, Lima Beans, and Peppers, had better be delayed to the last week in May. The first produce of the spring plantings will now be ready for use. Lettuce or Radishes, planted in cold frames in March, are matured from 5th to 20th May, and if covered up by straw mats at night, ten days earlier. In warm situations, on rich, light soils, the Radishes, Lettuce, Turnips, or Peas, planted in March, are fit for market. Rhubarb and Asparagus are also fit to be gathered, on early soils, the latter part of the month.

Additional labor is now beginning to be required, the marketing of crops occupying a large portion of the time, while the thinning out of sown crops, and the keeping down of weeds which are now showing themselves everywhere, entails an amount of labor not before necessary. To withhold labor at this critical time, is short-sighted economy, whether by the owner of a private or market garden; for let the crops planted and sown, once get enveloped by weeds, it will often cost more in labor to clean the crop, than it will sell for; it is not at all an uncommon occurrence to see acres of Carrots or Parsnips plowed down, after being carefully manured and sown, from neglect or inability of the owner to procure labor at the proper time. The rapid development of weeds is, to the inexperienced, very deceptive; a crop of Carrots, Parsnips, Beets, or Onions, may appear to be easily manageable at a given day in May; but a few days of continued rain occurs, and the crop, that could have been profitably cultivated on the 15th, is hopelessly over-grown on the 25th.

JUNE is one of the months in which we reap the reward of our operations in the market garden; at this time, the bulk of all the early crops matures. So far, nearly all has been outlay; now we receive the returns. In this district, our early crops of Asparagus, Beets, Cauliflower, Cabbage, Lettuce, Onion, Peas, Radishes, Rhubarb, Spinach, and Turnip, are sold off, and the ground plowed for the second crop, (except in the cases of Asparagus and Rhubarb), by the end of the month. For private gardens, succession crops of Beets, Bush Beans, Cabbages, Cucumbers, Lettuce, Peas, Radishes, and Potatoes, may still be planted, but it would hardly be profitable for market purposes; as it would occupy the land wanted by the market gardener for his second crop, besides the market buyer of the cities will hardly touch a vegetable or fruit behind its season at any price. He will pay 10 cents per bunch for Radishes in May, and will pass by a far better article of the same kind in July or August, though offered at one-fifth the price. He will give 50 cents per quart for Tomatoes, (half-ripe), in June, that he could not be induced to touch in October, if he could buy them at 25 cents per bushel.

The Cucumbers, planted in cold frames and forcing pits, are also marketable in the latter part of this month. Great care must be taken to have them abundantly watered in dry weather; inattention to watering, (particularly of all vegetables under glass), is sure to entail loss on the cultivator, by giving an imperfect or partial crop. Watering had better be done in the evening, whenever the surface appears dry, not by a mere sprinkling, but by a thorough soaking; not less than a gallon to every square yard of surface. As soon as the Cucumbers are all cut from the frames, the sashes should be piled up at the ends of each section, and covered with a shutter, and a weight of some kind put on the top, to prevent them being blown off by high winds.

July.—The remaining part of the spring crops are cleared off in the early part of this month, and by the middle of it, unless the season is unusually dry, all the ground is planted with the second crops of Celery, Sage, Thyme, Late Cabbage, Broccoli, Cauliflower, or Leeks. Little is done to these crops this month, as but little growth is made during the hot dry weather, and newly planted crops are merely stirred between the rows with the hoe or cultivator. Some of the other later crops are now maturing for market. Bush Beans, Cucumbers, Potatoes, Squashes, and in early places, Tomatoes; also succession crops of Peas, Beets, Onions, Cabbages, etc., such of these as only mature during the end of the month, render the second crops rather late, unless for the later crops of Celery and Spinach.

AUGUST.—Except the months of January and February, August is a month requiring less labor in the market garden than any other; usually all the planting has been done in July, and the long drouths common at this season, stagnate the growth of even our most luxuriant weeds, so that in this month, of all others, the garden ought to be clean.

Late plantings of Celery may be made, to the middle of the month, and still make fair-sized roots for winter. Spinach may also be sown for an early crop, to be cut off in fall. Ruta Baga Turnips should be sown early in the month, and the white and yellow varieties during the later part. If the fly attacks them, it may be kept down, so as to do but little harm, by frequent applications of lime, dusted lightly over the rows. Bush Beans and Peas, may still be sown for late crops. The Onion crop will ripen off during this month, and when convenient to market, should be offered for sale as soon as gathered, as the price received for those first sold, is frequently double that of those coming in ten days later.

SEPTEMBER.—The cool nights and moist atmosphere of this month begin to tell strikingly on the crops planted for fall use; Celery, Cabbage, and Cauliflower, now grow rapidly, and require repeated stirring of the soil with the plow, cultivator, or hoe. Celery, that is wanted for use towards the latter part of the month, may now be "handled," or straightened up, and the earth drawn to it by the hoe; in a week or so after, it may be "banked up" by the spade to half its hight, allowed to grow for another week or more, until it lengthens out a little further, when the banking should be continued as high as its top. In ten days, (at this season), when thus finished, it is blanched sufficiently to use, and should then be used, or it will soon spoil. Care must be taken that no more is banked up than can be sold or used, as it is not only labor lost, but is decidedly hurtful to the Celery, by making it hollow. The practice recommended by most authorities, and still practiced by private gardeners, is, to keep earthing it up every two weeks from the time it begins to grow; this is utter nonsense, resulting in giving Celery tough, stringy, and rusty—utterly unfit to eat, while the expenditure in labor would be twice more than the price it would bring if sold; for further information on this important subject, see article on Celery. The seeds of Cauliflower, Cabbage, and Lettuce, should be sown this month, from the 10th to the 20th, for the purpose of being pricked out in cold frames to be wintered over; it is very important that the sowing should be done as near these dates as possible, for if sown much before the 10th, the plants may run up to seed when planted out in spring, if much later than the 20th, they would be too weak to be wintered over. Shallots and Onions should also be planted this month, and Spinach and German Greens, or "Sprouts," sown to be wintered over, all now for spring use

OCTOBER.—This month corresponds in part to June of the summer months, being that in which the returns from the second crops come in. Celery, that has been banked or earthed up, now sells freely and in considerable quantities; all the crop should this month be "handled," and as much as possible earthed up. Cauliflower is always scarce and dear in the early part of this month, but unless the fall has been unusually moist, is generally not matured until towards the end of the month. Thyme, Sage, and all Sweet Herbs, should now be sold, from the beginning of the month, cutting out only every alternate row, as it gives the crop time to grow, so that the remaining rows spread sufficiently to fill the space. (See article on Thyme, etc).

The crops planted or sown last month, must now be carefully hoed, and the weeds removed; for, though weeds are not quite so numerous in variety as in summer, Chickweed, now very abundant, is one of the most expensive weeds of the garden to eradicate.

The plants of Cabbage, Cauliflower, and Lettuce, recommended to be sown last month, are now fit to be pricked out in the cold frames. (See detail of the process.)

NOVEMBER.—This month warns us that winter is approaching, and preparations should be carefully made towards securing all products of the garden that are perishable by frost. The process of putting away the Celery crop in trenches for winter use, (see Celery article), should be begun about the 5th or 10th of the month in dry weather; that put in trenches then, will be blanched sufficiently for use in six or eight weeks, but when sufficient help can be obtained, it will always pay well to bank or earth up a large portion of Celery by the spade, clear to the top; this will keep it safe from injury from any frost that we have in this month, and thus protected, it need not be put away into winter quarters—the trenches —before the end of November; put away thus late, it will keep without the loss of a root until March or April, when it is always scarce and high in price.

The great difficulty most persons have, is from stowing it away and covering it up too early; this practice of earthing it up to the top roughly in November we have only practiced for the past two seasons, but find the extra labor well repaid, as we are enabled thus to save this very valuable crop without loss. There is rarely need of applying any covering of leaves or litter to the trenches this month, and it cannot be too often told that the covering up of vegetables of all kinds in winter quarters should be delayed to the very last moment that it is safe to do so. Beets, Carrots, Cabbages, and Cauliflowers, must be dug up, and secured this month in the manner recommended in "Preserving Vegetables in Winter." Horseradish, Salsify, and Parsnips, being entirely hardy, and frost proof, need not necessarily be dug, although from the danger of their being frozen in the ground next month, if time will permit, the work had better be progressing.

All clear ground should be dug or plowed, and properly leveled, so that on the opening of spring operations can be begun with as little delay as possible. If draining is required, this is the most convenient time to do it, the ground being clear, and not yet much frozen.

Towards the end of the month, the sashes should be put on the Cabbage and Lettuce plants in cold nights, but on no account should they be kept on in day time, as it is of the utmost importance that they be not made tender at this time by being "drawn" under the sashes. I may again repeat that these plants are half hardy, and it is killing them with kindness to protect them from *slight* freezing. Cabbage and Lettuce plants may be exposed in any place without glass, or other protection, where the thermometer runs no lower than 10 above zero. Rhubarb and Asparagus beds will be benefited by a covering of 4 or 6 inches of rough manure, or any other litter, to prevent the severity of the frost; the crop from beds, thus covered, will come in a few days earlier, and will be stronger than if left unprotected.

December. — Occasionally, we have the ground open so that digging and plowing can be done to nearly the end of the month, but it is not safe to calculate much after the first week; though by covering up the roots, still undug, with their own leaves or with litter, we are often enabled to dig our Horseradish or Parsnips very late in the month, and like all other vegetables, the later they remain in the soil they grow in, the finer is the quality.

Celery trenches should receive the first covering, early in the month, if the weather has been such that it has been unnecessary before; the covering should not be less than 4 or 5 inches of litter or leaves, only taking care that the material is light, weight or closeness would prevent evaporation too much at this season, while the weather is not yet severe; the final covering should not be later than the end of the month.

The crops of Spinach, Kale, Onions, Shallots, etc., that have been planted or sown in September, should be covered up with hay or straw if their position is much exposed; if not, there is no particular necessity. When all has been secured safely in winter quarters, attention must be energetically turned to procuring manure, muck, and all available kinds of fertilizers; there is little danger of spending toc much in this way if you have it to spend—depend upon it, there is no better investment if you are working your Garden for Profit.

Lightning Source UK Ltd.
Milton Keynes UK
UKHW050214210223
417160UK00023B/290